A County - Lover's G

A Country-Lover's Guide to Wildlife

A Country-Lover's

Mammals, Amphibians, and Reptiles

KENNETH A. CHAMBERS

with illustrations by H. WAYNE TRIMM

THE JOHNS HOPKINS UNIVERSITY PRESS
BALTIMORE AND LONDON

Guide to Wildlife

of the Northeastern United States

This book has been brought to publication with the generous
assistance of the Anne S. Richardson Fund.

Manufactured in the United States of America

The Johns Hopkins University Press, Baltimore, Maryland 21218
The Johns Hopkins Press Ltd., London

Library of Congress Catalog Card Number 79–4338
ISBN 0–8018–2207–6

Library of Congress Cataloging in Publication data
will be found on the last printed page of this book.

To my Ann

Contents

Reptiles: Victims of Misunderstanding   **57**

Plates

Preface

If you have ever heard a frog chorus at dusk on a spring evening and paused to listen and wonder about it, if you are curious about the snake you sometimes see in your garden, if you are not sure whether that actually was an opossum you saw disappearing into the hedgerow early the other morning— if, in other words, your eyes are opening to the natural scene— then this guide should help you. It is designed to be an introduction to one aspect of rural life that can add immeasurably to your enjoyment of the countryside.

During my more than twenty-five years of teaching and lecturing about wildlife it has many times been suggested to me that I should write a book on the subject—a book not only for naturalists, but also for people who merely wish to learn a little about the animals they may see from time to time, a ready reference for hikers and campers, students and teachers, city dwellers who have homes in the country and people who live there all the time. Here, I hope, is just such a book.

I have purposely omitted the birds. They are certainly the most evident form of animal life, but there are already available many fine books about them. On the other hand, there are few books concerned with mammals, amphibians, and reptiles; in fact, as far as I know, this is the first book that includes those of the Northeast in a single volume. I have selected the northeastern United States because that is where I live, some of the time in New York City, where I work, but as much as possible at my farm near New Lebanon, New York.

In researching some of the material for this book I have received help from a number of people, notably from colleagues at The American Museum of Natural History. Among these I should like to thank Dr. Richard G. Van Gelder and Mr. Jean Augustin of the Department of Mammalogy, and Dr. Richard G. Zweifel, Dr. Charles J. Cole, and Mr. George W.

Foley of the Department of Herpetology. My special thanks go to my good friend Wayne Trimm, whose beautiful illustrations form an important part of the book. Most of all, however, I am indebted to my wife, Ann. Her constant encouragement and her constructive criticisms and suggestions have been of incalculable value. Her patience has been phenomenal! Finally, let me pay tribute to my lifelong friend "Deg" Firth, who, many years ago, opened *my* eyes to the natural scene.

A Country-Lover's Guide to Wildlife

Introduction

Howbourne

A winter scene at Howbourne

WE FIRST SAW our farm in late autumn, when the fall colors were past their peak and had become drab and lifeless. It had been raining, and above the sodden fields the sky was still gray and threatening. Wet leaves whirled along the hedgerows as a cold wind tore them from trees already half-stripped of their foliage.

In the center of a treeless, sloping field the stone house sat starkly. A tangle of brambles and dead, yellow-brown bedstraw engulfed it, and the windows and doors were boarded up. A crow flapped heavily overhead, rising and falling as it battled the wind, but there was no other visible sign of animal life, no sound but the moaning of the wind in the eaves and a rustling and fluttering as it beat against the weeds. Everywhere were signs of indifference and neglect. It was a sad and lonely landscape.

Today, some fifteen years later, the scene has changed. How-bourne, as we have named the farm, seems a happier place. A chorus of birds welcomes the morning, and frogs lift their voices to the night. Muskrats forage in the wet bottomlands and raccoons hunt along the hillside. At the lawn's edges rabbits and woodchucks nibble contentedly, and ruffed grouse strut amid the beebalm and wild thyme under the cherry trees.

Even winter seems friendlier, and although often the wind howls and blows the snow into deep drifts against the house, now we have evergreens to blunt its force and give us shelter. There are days when it is still and peaceful, when the chicka-dees at the suet feeders call back and forth to each other, and when we discover the tracks of deer and squirrels, mice and rabbits crisscrossing the snow near the house.

This transformation took time and work. Undoubtedly some animals were always present, but we were not aware of them at first. There was scant food or shelter for them around the house, and the nearby open fields offered little more. One of our early projects was to fill some of the barren fields with conifers: Scotch pines where the soil was poor and thin, and white spruce, white pine, red pine, and Austrian black pine in more fertile spots. These trees have grown, and with their yearlong cloaks of soft green they relieve the harshness of the winter scene and provide a haven for wildlife. We have planted autumn olive, multiflora rose, honeysuckle, lespedeza, and high-bush cranberry. They are beautiful and need no care. They are also hiding places for animals, and nourish them with their fruits. Around the house there are now flower beds, and lawns dotted with shrubs and trees. These have attracted an ever growing assemblage of birds, and many of them nest there. On warm summer evenings the crab apple trees sparkle with fireflies and bats flutter and dive above the shrubberies.

To add an aquatic habitat we built a pond. It lies within a hundred yards or so of the house, in a natural basin surrounded by brush. Frogs and salamanders breed there, and turtles bask on its banks. Many animals visit it to drink, and sometimes migrating ducks and shorebirds pause to rest and feed there.

Howbourne has come to life again.

I spend a great deal of time watching the animals that live around us. Many of the wildlife observations I have made appear in the following pages, and most of the essays are based on incidents I have witnessed there. What I see is not always spectacular or of a pleasant nature, but it is real and uncontrived. It holds an enchantment of a very special kind.

One of the greatest of all country-lovers was the renowned author-naturalist, John Burroughs. In the preface to his book *Boy and Man* he states, "The most precious things of life are near at hand, without money and without price. Each of you has the whole wealth of the universe at your very door." I think that says it all.

Area Covered by This Book

What is the Northeast? It could encompass a very large area indeed, but for the purpose of this book I have defined it as being composed of Maine, New Hampshire, Vermont, Massachusetts, Rhode Island, Connecticut, New York, extreme eastern Pennsylvania, and New Jersey north of the Raritan River. This seems to me still to constitute a sizeable chunk of country—one in which the animal life is diverse enough to be interesting and yet fairly separable from that of the surrounding areas.

How to Use This Book

All of the animals covered are to be found within the Northeast, but few of them occur throughout the entire area. The distribution tables in appendix A indicate which of the Northeastern states are inhabited by each species. (Bear in mind, however, that it is possible that you may chance upon an animal outside of the range shown by these tables. Animals are mobile. If the population of a species increases radically or if the habitat available to it shrinks in size, individuals at the borders of the range may be forced to seek greener pastures. Eventually such testing of new areas might lead to a permanent extension of the range.)

Each of the three main divisions of the book—amphibians, reptiles, and mammals—is introduced by some basic information on the group as a whole. This material is the real foundation for learning about the animals that come later. Following these introductions, the groups are split into major subgroups. (For example, reptiles are split into turtles, snakes, and lizards.) At the beginning of each of these sections is a brief essay on one of the animals that will follow. This shows some environmental relationships. The essay precedes a summary of the general characteristics of all of the animals within the subgroup, together with some other facts about them. Then come the descriptions and life stories of the various species.

I have selected the more common animals for greater attention. These, after all, will be more frequently seen, and presumably will be the animals that most people would, therefore, wish to know about. The less common animals are there too, but they are dealt with in less detail. Near the back of the book is a section entitled "Animal Relationships." This explains the basic system used by biologists for classifying animals. Also near the back of the book is a glossary that defines some of the technical terms used. Finally, there is a brief listing of other publications that present additional information about the animals discussed.

Observing Animals

In order to observe animals in the wild there are several factors that one needs to remember. Probably among the most important of these is that it takes patience—patience to find them in the first place and then more patience to keep them under observation without disturbing them. Obviously, the more noise one makes, the less is the chance of accomplishing either of these objectives. People who go rushing around stand very little chance of seeing much in the way of wildlife. It is far better to concentrate on creating as little disturbance as possible. In some cases I find that just selecting a comfortable, secluded spot and sitting there quietly for a while yields surprisingly good results. One never knows what may turn up. It helps, of course, if one can decide on a promising spot in which to do one's sitting. This is something that comes with experience. If you consistently see fresh deer tracks in one spot at the edge of a pond it is reasonable to assume that deer come there regularly to drink. Late evening or early morning would be the best time to see them. If there are several shed snake skins on or around a certain pile of rocks, the chances are that their erstwhile owners may be found living there, and that on sunny days they may bask nearby. If there are little, well-worn paths running through the grass at the edge of a field, then some small creature—a mouse perhaps, or a chipmunk—has made them and is probably still using them. These are the kinds of places near which one should hide.

Sitting still and silent is one way of seeing wildlife. Another is to walk slowly along, keeping your eyes open. Although there is usually something to be seen at any hour, I have found that evening walks and "dawn patrols" are the most rewarding. Be ready to "freeze" at the slightest movement or sound. If you have not already been spotted, and remain motionless, animals will often approach quite closely, especially if you are wearing fairly drab-colored clothing. As I walk I often turn over rocks and rotting logs. This frequently produces interesting results in the form of salamanders, snakes, and other small animals. (Return such rocks and logs to their original positions; they may be the homes of animals that happen not to be there at the time.)

Remember that it is not always necessary to see an animal in order to know that it is around. Tracks, droppings, the remains of kills, freshly gnawed tree bark or newly clipped grasses, calls, nests and dens—all are important evidence of the presence of wildlife in one form or another. Once such evidence has been noted it is often a matter of using some of that patience mentioned earlier and of keeping a watchful eye on that spot. There are many methods used in finding certain particular species. Some of these are mentioned later on in this book; you will develop others from your own experiences.

Once you begin learning about animals by watching them it is a good idea to keep some kind of diary or log of your observations. Write up your notes either immediately or, at the latest, before you go to bed at the end of the day. It is amazing how little time it takes to forget much of what has been seen. Jot down the date and the weather conditions (including temperatures if you can), and list the animals you have seen and where you saw them. Add whatever details you think are important. All of this information will be valuable to you in future years. By consulting your notes you will be able to forecast, pretty accurately, many annual events: the time to expect the first frog chorus (and which kind of frog will be calling, and where), when the last woodchucks disappear to hibernate in the fall, and so on.

This is one value in taking notes: you build up a gradual compilation of knowledge of animals by your own efforts in the field, and not merely by reading a book or two. Who knows? You may observe something about an animal that nobody else has noticed! In addition to this accumulation of data, however, a good log book is the best way I know of recalling a walk in its entirety. Sometimes I will read over some of my notes (now going back almost forty years) and in the reading will relive days in the field as if they had occurred only yesterday. To me this is a most satisfying way to spend part of a long winter evening.

Man's Effects upon Wildlife

As human settlement in the Northeast steadily increased, so, too, did man's impact upon the landscape. Down through the years more and more forest land was cleared for agriculture and to meet the increasing demand for lumber and paper products. Animals living in this forest land were thus deprived of food and cover and places in which to raise their young. In hilly areas the roots of the trees had held the soil in place. Once the trees were gone much of the soil eroded and was washed downhill and into the valley streams. This caused the streams to become more shallow and, with a lack of foliage for shade, the water increased in temperature. Many of the aquatic plants and animals that needed cool water to survive died out, and this also affected larger animal life. In the course of urban development many marshes and other wetlands were drained.

This lowered the water table in those areas, and created more changes in vegetation.

Many animals were unable to adapt to these radical changes in the landscape. Overexploitation of other animals for furs, food, and sport, and more recently the overuse of pesticides, have all taken their toll. Some fur-bearers, including the marten, fisher, otter, and lynx, are now rare. The bog turtle is found only in a few scattered localities. Some snakes, such as the hognose snake and the massasauga, are likewise far less common than they used to be. The elk, timber wolf, wolverine, and mountain lion are among those mammals that can no longer be found anywhere in the Northeast, and with the possible exception of the mountain lion there seems little chance that we shall ever see them return.

When an animal's habitat is changed or destroyed it has only two alternatives: it can adjust to the new situation or it can move to another, similar area. If it is unable to do either of these it must inevitably die.

To those of us interested in animal life all of this is very sad, even though we realize that many of these reductions in wildlife were probably unavoidable. The land can support only so much, and once the human population had increased beyond a certain point other forms of life inevitably decreased, or died out altogether.

Yet the picture is not completely gloomy. Controls on hunting, trapping, and fishing, new antierosion techniques, a wiser use of pesticides, the establishment of parks and wildlife refuges, and many other wildlife-oriented practices seem to be gradually improving the situation in many areas. Most of the present Northeastern wildlife species now appear to be holding their own; some are certainly increasing in number. While still rare, the fisher is beginning to turn up in places from which it had long since disappeared; beavers are again quite common after being driven almost into extinction; terrapins are on the upswing in some coastal areas. Certainly we shall never again see the abundance and variety of wildlife that was present at the time of the early settlement of the Northeast, but at least we now seem to be trying to maintain what we have. There are still many animals to be seen, and there is still a wealth of pleasure and excitement to be gained in seeing them, whether in some remote woodland or in one's own backyard.

These days more and more of us are feeling a need to relax from the pressure and strain of our modern, fast-moving way of life, and in growing numbers we are seeking quieter, more secluded places. There, if we are so inclined, we may discover the wild creatures. Once we develop some appreciation for them and for their ways of life, our own lives can take on entirely new meaning. Perhaps, after all, this is the greatest value that wildlife has to offer us.

Amphibians

Dwellers in Damp Places

The Howbourne pond

THE WORD *amphibian* is applied to animals that typically begin their lives in water and later move out onto the land. *Amphibious* has a different meaning. It can be used to describe many animals that spend much time in the water as well as ashore. Among these would be numbered water snakes, turtles, seals, beavers, and otters.

An amphibian differs from other groups of vertebrates in several respects. It has a smooth or granular skin that may have specialized glands. Many of these glands secrete a mucous substance that keeps the skin moist. This is why most amphibians feel wet or slimy. The mucus not only prevents the animal from becoming dessicated, but it also protects it from infection. So it is best to have wet hands if one wishes to handle an amphibian.

Amphibians usually lay their eggs in water. The eggs develop into larvae that at first have feathery, external gills. These gills have the same function as the gills of a fish—they extract oxygen from the water. Later the gills disappear, to be replaced by lungs. From then on the amphibian breathes air. There are exceptions to all of this, but this is the typical situation.

Among the land vertebrates, amphibians share with reptiles the condition of cold-bloodedness. This simply means that their body temperature varies constantly. Just how much it varies depends on the temperature of their environment. If a cold amphibian happens to sit on a warm rock, the temperature of the amphibian will rise. If a warm amphibian leaps into cold water, the reverse happens and its temperature drops. This, of course, is in contrast to warm-blooded animals such as ourselves, where the body temperature remains almost constant.

The northeastern section of the United States can become very cold in winter. Often the air temperature drops to below freezing. This means that if an amphibian (or a reptile) were subjected to such temperatures it would itself freeze—and die. Obviously, most of them do not suffer this fate. They avoid it by moving into protected sites during the fall, and hibernating. These sites may be underground, back in rock crevices, or even underwater. They have one thing in common; they are beyond the frostline. So the hibernating animal does not freeze. Nor does it starve during its long winter sleep. Its entire metabolism slows down, so that the fat stored in its body is used up only very gradually. Hibernation is a wonderful mechanism, and this is merely the outline of a very complicated process.

So much for some of the characteristics of most amphibians. How about the actual animals? What amphibians do we have, here in the Northeast? There are three main groups of amphibians in the world. We have representatives of two of them: the salamanders, and the frogs and toads that are lumped together into one group. (The third group is composed of animals called caecilians. These are burrowing, wormlike animals found in some tropical areas of the world.)

Frogs and Toads

The Early Spring Pond

It is an evening in late March, and I have grown tired of sitting indoors. I pull on my boots and head toward the pond. The air is chill, but it is not the hard, numbing coldness of midwinter. Already there have been several days when the temperature has reached the mid-fifties, and there have even been days of prolonged sunshine. But now rain falls as a soft drizzle, and its warmth melts the last of the snow in the open spaces. Indeed, the long-frozen soil has lost its iron and yields in slight sogginess underfoot. Newly exposed brown grass hugs the ground, flattened by the weight of past snowfalls. Soon the first green shoots will shoulder aside the old, dead leaves, but as yet the grass shows no sign of life.

The pond at Howbourne is roughly oval in shape and sits in a basin on a shallow hillside. A tangle of leafless shrubs dotted with clumps of trees surrounds it and makes it difficult to see. Some paper-thin ice still edges it, but only at one end, and even this will soon be gone. The downhill side, where a large earthen dam holds back the water, curves smoothly, but the other banks have indentations forming miniature bays and headlands. At one end is a larger headland, where a graceful birch leans outward between two large boulders. There is a small, shallow cove next to it, lined with sedges that form tussocks at the water's edge. On the side nearest the birch there are brittle, broken cattails. Like other plants around the pond, the cattails are shaded brown, but here and there the fluffed-out remains of last year's seed heads add a flimsy touch of white. Small whisps of vapor have formed over the pond, and although it is still raining the rain itself is mist-fine and fading. Finally it stops completely.

The pond stretches serenely into the shadows and the mist. Sometimes the water flashes momentarily as it catches the light from a star, but there is no real movement. It does not lap at the rocks or wash along the shoreline. Like the shrubs and trees, sedges and cattails, it is still and silent.

Suddenly, from among the sedge tussocks, there comes a faint call. It is tremulous but distinct. It quavers out upon the quiet night for a brief moment and is gone. There is a long pause. The call is repeated, this time with a little more strength. And again. And again and again. Time passes, and gradually the owner of this lone voice seems to gain in confidence. The calls lose their quaver. They become louder, and resound urgently out across the pond at frequent intervals.

Like an echo, another voice joins the first. And another. From all around the shoreline, from the shrubs and sedges, more high-pitched voices are added to the growing chorus. The air pulsates with piping calls. They drive through the mist, shrill to the stars.

For an hour or so the calling continues, but as the temperature drops the chorus loses its strength. One by one the voices are stilled. There is silence again. The stars shine down upon the pond and the mist. A puff of wind ruffles the water, and the thin ice at the pond's margin lifts a little to the ripples.

This brief evening chorus would go unnoticed by most people. Yet, to the initiated, it is an exciting sound. It is the sound of spring peepers, newly awakened from hibernation. In the northeast this calling is one of the first signs of spring in the animal world. Later in the season, when the nights are less cold, the choruses are greater and may continue past midnight. After a warm rain I sometimes take a flashlight and walk down to the pond. Drops of water fall from new leaves. The fresh grass is squelchy as I walk.

Still several hundred yards from the pond, I can hear the peeper chorus. At this range it is a restful sound, rising and falling like the soft wash of wavelets on a shingle beach. The sound melds with the night, like the chirruping of crickets, so that it almost goes unnoticed. It is a sound to sleep by, a sibilance of whispers.

As I approach the pond the sound becomes less soporific. The volume grows. The chorus becomes more strident. Waves of shrill calls beat back and forth across the black water. They thrill through the night air and drown the senses with their magnitude. I pause and listen. Below me the wet reeds rise, blurred clumps on the shoreline. Flat clouds drift across the dark sky.

I inch closer to the water's edge. As the frogs sense an intruder, the voices of those closest to me die away, but on the far side of the pond they continue to call. I squat down on my heels and imitate the call. Again and again I call. Suddenly there is an answer from nearby. Other frogs join in, and soon the chorus swells to its original volume. It dins upon my eardrums. I locate a voice about four feet from me and aim the light. When I am ready, I switch it on. In its blinding glare I see nothing at first. I move the beam slowly left and right, directing it at the base of the tussocks, where they meet the water. A quick glitter—and there he is. His head is raised. His forelegs are straight and vertical beneath his small body, half-hidden by the leaves of the sedge. The light reflects from his large, transluscent vocal sac. It swells from beneath his chin like a pearly miniature balloon. As I watch, it deflates and disappears. The peeper shifts position slightly. It moves the rear of its body with several quick, jerky motions. I see the X-shaped marking on its back. Now its throat begins to vibrate. There are a few partial inflations and then the sac swells forth again. When fully expanded it is half as big as the peeper's body. He recommences his calling. The light that bathes him does not seem to bother him. He winks and blinks a little, but his vocal sac continues to pulsate with passion.

From the far end of the pond I hear a different sound. It is much lower pitched than the peepers' calls. It sounds vaguely like ducks quacking, but is more gutteral: a continuous "wark-wark-wark-wark-wark." It is a croaking sound, the kind many people wrongly associate with all frogs.

I cannot see the frogs that make this sound, but I know they are wood frogs. They are larger than peepers, brownish or pinkish with a black mask around the eyes. They often call even before the peepers, but their voices are not as musical. They call only for a week or so and then are gone from the breeding pond until next year. Once in a while a lone wood frog may be spotted as it hops across a shaded woodland path. Once in a while a wood frog rustles dead leaves as it moves over the forest floor. But these are chance encounters.

Toward the end of April the American toads breed. They have loud, musical trills, and often breed during the day as well as after dark. They are followed by other species, each calling more or less at its appointed time, though there is much overlapping. One of the last to begin is the bullfrog, whose deep bellowing is held by some to sound like "jug-er-rum." Here is a monster among frogs: a green-snouted, brown-bodied, cream-bellied basso profundo whose head and body may reach a length of eight inches and whose cavernous mouth will engulf anything it can catch and swallow, including its own kind.

Most frogs and toads must return to the water each year to reproduce. Thus the choruses. The calls are made by the males trying to attract mates down to the pond; they are serenades. The males rarely call at other times, and then only in half-hearted fashion. It is the urge to breed that is the real stimulant.

For the most part, frogs and toads are innocuous. They seem entirely undramatic. Yet their lives are dramatic enough. Those that succeed in reaching adulthood do so only after over-coming tremendous odds. If the eggs from which they originate are laid at the wrong time or in the wrong location or at the wrong depth, if drought removes the water, or flood deposits them on land, they do not hatch. If they do hatch, the re-sultant tadpoles are preyed upon by fishes, aquatic insects, crayfish, birds, and almost anything else capable of catching them. The survivors eventually transform into miniature frogs and toads. They struggle ashore to continue their growth. But the slaughter goes on. They still face a host of enemies. Many kinds of snakes and birds and mammals harvest them. Barely enough escape to reach an age at which they are able to breed and perpetuate their species.

Those that succeed are wary. They are normally seen only accidentally, often only fleetingly. Although some species are brightly colored and boldly patterned, they blend almost magi-cally into their backgrounds. Unless they move they are often overlooked. When they do move it is usually to present only a quick glimpse as they plop into the water or vanish into the long grass. Ecologically they are important links in the food chains of other animals. But to those with more than a mere scientific interest in them they represent far more. Their pres-ence adds color to the morning. Their voices bring the nights to life.

There are about 180 species of frogs and toads living today. These comprise the order Anura. More than 70 species are to be found in the United States and Canada, and of this number 13 occur in the Northeast.

As adults, all Northeastern frogs and toads are tailless. While most members of the group are smooth skinned, many toads have skins that are warty, and some frogs, such as the gray tree frog, have granular skins. Toads have no teeth, but frogs have minute teeth in their upper jaws, though they are so small that their presence can be felt only as a faint raspiness.

Behind each eye, toads have a fairly large brown lump. These are the parotoid glands. They secrete a poisonous substance strong enough and distasteful enough to make a dog or other predator drop the toad if it takes one into its mouth. (The secretion is not harmful to humans when handling a toad.)

There is some variation in the feet of frogs and toads. The so-called "true" frogs have webbing between the toes to assist them in swimming. The amount of this webbing varies with the species. Tree frogs have a small suction disc at the tip of each finger and toe, enabling them to stick to leaves, twigs, or whatever they happen to climb upon. The hind feet of toads have brown, horny "tubercles" that help them to burrow.

In general, breeding in frogs and toads takes place in spring and early summer. The males are usually the first to arrive at the breeding pond, and they begin calling to attract females. When a female approaches, a male clambers onto her back and grasps her with his front legs. The eggs are usually released soon afterward by the female, and as they are extruded into the water the male releases sperm over them. This is called external fertilization. Some species of frogs deposit their eggs singly; others deposit them in masses. Toads lay their eggs in long strings. In both frogs and toads, the eggs are surrounded by a transparent, jellylike envelope that absorbs water and swells out around them.

In anywhere from a few days to several weeks, depending upon the species, water temperature, etc., the eggs hatch into larvae (tadpoles). The larvae have tails and external gills, but these gills are later replaced by internal gills. At first the tadpoles are vegetarians, rasping at the stems and leaves of water plants with thick, horny lips. In time the hind limbs appear, and once they have attained a good size they are followed by the front limbs. Now the tail begins to shrink as the tadpole gradually absorbs it. The gills vanish and lungs are used for breathing air. The eyelids form, and the long, vegetarian-type intestine shortens for a carnivorous diet. Finally, the young frog leaves the water, perhaps with the vestige of its tail for a short time. While most species transform from tadpole into frog by their first winter, here in the Northeast the green frog spends one winter as a tadpole, and the bullfrog two winters.

All frogs and toads, as adults, are carnivorous. They feed on a variety of insects and other small animal life such as earthworms, spiders, centipedes, and small fishes. Many use their sticky tongues to catch their prey. Larger species such as the bullfrog may be cannibalistic. Since all species eat the same types of food, there is no separate food section given under any of the accounts of individual species that follow.

Although active mostly at night, they may also be seen during the day. They are, however, rather secretive. Even those that may sit in the shallows of ponds or on the banks near the water are generally concealed by sedges or other plants. They become less shy during the breeding season. This is when the males call. The calls are made by vibrating air between the lungs and vocal sacs. The latter are external in some species and internal in others. Females may sometimes squawk if caught by a predator or if roughly handled, but they are essentially voiceless.

All frogs and toads in the Northeast hibernate in moist locations where freezing temperatures do not reach them.

These creatures help to reduce the populations of some garden pests and are useful to man in several other ways. The leopard frog, in particular, is used for dissection exercises and studies in anatomy by biology classes, and males are often used in human pregnancy tests. The legs of some of the larger species are used as human food.

Frog catching a fly

Ecologically, frogs and toads are important food items for other animals, which may feed on tadpoles or adults or both. Among these predators are many species of fish, several species of snakes, wading birds such as herons, some hawks and owls, raccoons, skunks, foxes, weasels, and otters.

There is much folklore regarding frogs and toads. Many people still believe that handling toads will give them warts. Some people used to think that a toad had a ruby inside its head. Neither of these beliefs is true, nor is the idea that frogs and toads come to earth with the rain. (Sometimes enormous numbers of young frogs or toads leave the water during a rainstorm and may be seen everywhere for a brief time. However, they leave the water just as often in dry weather.) Another story that continues to crop up from time to time is that swallows turn into frogs or toads in the fall and hibernate in ponds and wells. The basis for this myth is that in pursuing flying insects, swallows may be seen skimming low over a well or a pond. If this is observed in the fall, when these birds are migrating, it seems to add weight to this story. One belief that has more foundation is that some frogs give warning of when it is going to rain. It is true that an increase in humidity often stimulates some species to call briefly outside of their breeding season—the gray tree frog is an example—but other factors, such as gusts of wind, may cause the same reaction.

One thing is certain, no Northeastern frog or toad represents any danger to man. All can be handled with impunity.

Keeping Frogs and Toads in Captivity

A five-gallon aquarium tank can serve as a suitable container. It will require a lid, and you should be sure that, in making air holes, you punch from the inside of the lid; otherwise, a jumping frog may injure itself on the ragged edges of the holes.

For toads, tree frogs, wood frogs, and other species that spend most of their time on land, a terrarium type of environment is required. Line the bottom of the tank with pebbles or crushed rock (for good drainage) and then put in a layer of soil about three inches deep. Half-bury a couple of larger rocks or pieces of wood in the soil and, to make things more attractive, plant some small ferns or grasses. You might even try putting in a few clumps of moss. Keep a shallow container of water in the tank so that the frogs can soak if they wish. Change this water frequently—daily if possible, but at least three times each week. If you are using city tap water, let it stand overnight before putting it into the container. This will get rid of any chlorine that may have been added to the water and that might injure your frogs.

For green frogs, bullfrogs, and other more aquatic species you need do nothing more than half-fill the aquarium tank with water and then arrange several large rocks so that the

Gray treefrog on a window pane

frogs can get out of the water if they wish. Again, if using city water, it is a good idea to keep a couple of lidless gallon jars filled with water standing nearby to evaporate any chlorine. Change the water at least once a week.

When feeding, remove individual frogs and feed them in another container. This will help keep your tank clean. Feed frogs and toads earthworms, pieces of raw meat, or raw fish. Mealworms can be used, but not as a steady diet. Hold the food item with a pair of forceps, and touch it gently to the corner of the animal's mouth. Wiggle it slowly to make the frog think it is alive. At first it may take a little time for any reaction but frogs and toads soon learn to grab for food presented to them in this way. Feed them every other day, and each time give them all they will take.

One easy way to feed tree frogs is to turn on the lights in a room after it gets dark and to then place the frogs outside on a windowpane. The light attracts flying insects, and the frogs use their suction discs to move around on the glass and catch the insects. (Stand nearby during the entire process, or the frogs will escape.)

Try to change your frogs and toads from time to time, and if possible release your earlier specimens in the same area where you originally captured them.

Eastern Spadefoot Toad *(Scaphiopus holbrooki)* PLATE 1

Adult Size. 1.75″ to 2.5″

Spadefoot toad

Description. This is a squat and stocky-bodied animal with a fairly smooth skin. Warts are present, but not as easily discernible as on the American or Fowler's toad. The eastern spadefoot's eyes are large and protruding. The name comes from the single, dark-colored, horny, sickle-shaped structure—the spade—on each foot. This assists in burrowing. The upper parts of the toad are usually brownish, but may be gray or black. There is a yellow or greenish yellow stripe running down the body from behind each eye. Each stripe curves in toward the center of the body, then out toward the side, and then back in again at the rear of the back. The two stripes thus form a vaguely hourglass shape. There is usually a similarly colored stripe down each side of the body. The underparts of the toad are whitish, but gray toward the rear.

Male's Call. The male toad utters a loud, harsh "warrk-warrk-warrk," usually while it is lying at the surface of the water, although its call can sometimes be heard coming from below ground. When a chorus is under way, the sound is like that of a flock of crows.

Breeding. In the Northeast, this species usually breeds during and after warm spring rains. The eggs are laid in irregularly shaped, elongated packets up to twelve inches long and about one inch wide. They hatch very quickly—usually within two days—and the tadpoles transform into toads in two to eight weeks.

Habitat. Since this is essentially a burrowing animal, it is found where the soil is soft or sandy. It emerges from its burrow only after dark and breeds only after heavy rain, often in shallow, temporary pools.

Range in the Northeast. Southern New England, southeastern New York (mostly on Long Island), eastern Pennsylvania, and New Jersey

Similar Northeastern Species. Both the American and Fowler's toads have a prominent parotoid gland behind each eye, and wartier-looking skins.

American Toad *(Bufo a. americanus)* PLATE 1

Adult Size. 2″ to 3.5″

Description. Like most toads, this is a plump-looking animal with a short, wide head and a dry "warty" skin. Behind each eye, at the top of the shoulder, is a raised, smooth, light brown lump. This is the parotoid gland. Both color and pattern vary considerably. In general, the ground color is brown, but it varies from almost yellow to a deep reddish brown. There are often various-sized patches of lighter, yellowish skin, and there are always patches of black skin on the back, sides, and legs. Within the *largest* of these black patches are *one* or *two* large warts. (The smaller black skin patches may have more than this.) Sometimes there is a light stripe running down the center of the back. Most of the underside is grayish, usually with dark spots in the chest region. The throat is dark.

Male's Call. The call is a sustained (up to thirty seconds) and quite musical trill, often falling off at the end of the trill.

Breeding. This species breeds mostly toward the end of April, although I have heard some males calling during the first week of April and others on into early July. It breeds and calls by night or day. The eggs number up to 8,000 and are laid in a single row in two long *strings* of jelly. These strings are usually no more than 0.25″ in diameter and may be piled back and forth on top of each other in the shallows or wound around the stems of plants underwater. There are two jelly envelopes to each string. The eggs hatch into tadpoles in three days to almost two weeks, and the tadpoles transform into tiny toads in about eight weeks. (I once found many newly transformed toads around a small pool on 30 May. This is unusually early.)

Habitat. The American toad is liable to turn up almost anywhere, from thick woodland to open fields, gardens, old foundations, barns, and even locations within cities. During the breeding season it is found in the shallows of ponds, marshes, ditches, and temporary pools.

Comments. The American toad often burrows into soft soil. Like most toads, it moves in a succession of short hops (rather than by leaping, as is the case with the majority of frogs). If handled, it may secrete a liquid that can irritate the eyes if accidentally rubbed into them. It is a good idea to wash one's hands after handling this animal.

Range in the Northeast. Common throughout all of the Northeastern states

Similar Northeastern Species. Fowler's toad looks very similar to the American toad but has at least three warts in each of the larger black patches of skin. Unlike the American toad, it has no spots on the chest. Where their ranges overlap these two species may interbreed, resulting in hybrids that cause great confusion even to experts.

Fowler's Toad *(Bufo woodhousei fowleri)*   PLATE 1

Adult Size. 2″ to 3″

Description. Fowler's toad is very much like the American toad, with a plump body and a short, wide head. The skin is warty, and prominent smooth, brownish parotoid glands are located at the top of the shoulder, behind the eye. The color of the toad varies greatly, except that there is always a cream stripe running down the center of the back. The upper parts are brownish, sometimes with a greenish tinge. There are dark skin patches present on the back, sides, and legs. Within the *largest* of these black patches, there are *at least* three warts and often many more than that. The underside is a light buff color. Males have a dark throat.

Eastern spadefoot toad

American toad

Fowler's toad

Northern cricket frog

Western chorus frog

New Jersey chorus frog

Gray treefrog

Spring peeper

Male's Call. The male gives out a loud, nasal "waaaaaaaaa" that lasts no longer than three or four seconds and that in no way can be described as a pleasant sound.

Breeding. From mid-April to the beginning of August, the Fowler's toad is to be found at the breeding sites. The female lays eggs in a long, jelly-enclosed string, sometimes in a double row within the jelly. Up to 8000 eggs can be found together in one of these jelly strings. Within a few days the eggs hatch. Tadpoles usually transform into adults in six to eight weeks.

Habitat. These toads live mostly near salt or fresh water, along beaches, in inland sandy areas, pine barrens, fields, and woods. They breed in the shallows of ponds, swamps, bogs, etc., or in ditches and slow-moving streams.

Range in the Northeast. Southern New Hampshire, throughout Massachusetts, Rhode Island, Connecticut, southeastern New York (including Long Island), and south into New Jersey

Similar Northeastern Species. The American toad differs from the Fowler's toad in that it has only one or two warts within each of the larger black skin patches. Hybrids are common between American and Fowler's toads where their ranges overlap.

Spring Peeper *(Hyla c. crucifer)* PLATE 1

Adult Size. 0.75" to 1.25"

Description. This small frog has a rather pointed snout, smooth skin, and a small suction disc at the tip of each finger and toe. The basic color of the upper parts is usually brown or gray, but an occasional frog may be almost olive. The most outstanding marking is a large X on the back, reaching from a little way behind the eyes almost to the groin. This cross is often not complete. It is darker than the ground color. Other similarly colored markings include a line between the eyes, a line from each eye running out along the snout, and some banding on the arms and legs. The underparts are buff to pale yellow, and males have bright yellow in the groin.

Male's Call. The male utters a single, clear, high-pitched piping note, rising toward the end and repeated at intervals of about a second. When the frog first begins its calling there is a slight bubbling or trilling within the note, but this tremulous quality disappears after the frog has been calling for a while. If many peepers are calling in chorus, the sound is almost deafening when heard at close quarters. This species often calls out of the breeding season,

Spring peeper calling

and I have heard it on into late October. At such times I have noticed that gusts of wind or aircraft passing overhead will often stimulate a faint chorus.

Breeding. Although these frogs breed mostly in April and early May, I have many times heard good choruses as early as March 18. The eggs are laid singly, usually on submerged vegetation near the bottom of the water. Each female deposits up to 1,000 eggs, and the resultant tadpoles transform into frogs during July.

Habitat. While breeding can occur in almost any fresh-water area, including temporary pools and ditches, this species is usually in greatest abundance in woodland ponds and swamps and in lowland marshy areas. During the breeding season the males almost always call from beneath grass tussocks or where plants afford good cover in or near the shallows at the edge of the water. Frogs moving toward the breeding sites may be heard calling from trees and shrubs. Once the breeding season is over they can occasionally be heard calling from sites in dense brushland or from quite high in trees.

Range in the Northeast. Common throughout the Northeast

Similar Northeastern Species. The New Jersey chorus frog, the western chorus frog, and the upland chorus frog all have striped backs (though not always complete stripes). They have a light line running along the upper lip. The northern cricket frog has a stripe running up its back and branching out to each eye. The upper surface of the snout is the same color as this stripe, which may be green, brown, red, gray, yellow, or white.

Gray Treefrog *(Hyla v. versicolor)* PLATE 1

Adult Size. 1.25″ to 2″

Description. This stocky-looking frog has a very rough, granular skin and a large suction disc at the tip of each finger and toe. The color of the upper parts varies. An individual frog may change from light gray through various shades of gray or brown to green. On the back there is a large dark marking shaped vaguely like a star, and below each eye is a light spot edged with a thin black line. Other markings include dark bars on the legs and toes. The underside is very granular and mostly whitish, except for the throat, which is darker. There is bright orange in the groin and on the underside of the hind limbs. None of this orange is visible when the frog is sitting.

Male's Call. The gray treefrog utters a loud, clattering trill usually lasting about two seconds. At the breeding ponds these calls are made from low shrubs, among clumps of sedges, from the shallows, or from mats of floating algae. I have often heard single frogs calling during the day, usually just before or during a rain, but also in sunny, dry weather. During July and August they frequently call from quite high in trees.

Breeding. This species breeds from late April through July in the Northeast, and on into August farther south. Although each female lays well over 1,000 eggs, they are laid in small clumps of up to 40 (but usually less than that), and are attached to vegetation at the surface of ponds and marshy pools. The eggs hatch into tadpoles in less than a week. These tadpoles are very easy to identify, for they have fairly bright orange crested tails speckled with black. The tadpoles transform into frogs in six to nine weeks.

Habitat. Except when breeding, these frogs are found mostly in trees, where their coloring blends perfectly with that of the bark. I have also found them on old stone walls and sometimes on the wooden walls of farm buildings.

Comments. Skin secretions from treefrogs may irritate one's eyes if inadvertently rubbed into them. Be sure to wash your hands after handling any of these creatures.

Not uncommonly, gray treefrogs may appear at night on the window screening of our house during the spring and summer months. Here they find a real bonanza in the form of insects attracted by lights within the house, and may spend some time climbing around on the screens, feeding on small moths and flies.

Range in the Northeast. Southern Maine and throughout the rest of the Northeast except for northern New Hampshire and northern Vermont

Similar Northeastern Species. There is really no other frog in the Northeast that can be confused with this species. It is considerably larger than any of the other Northeastern treefrogs, and the discs on fingers and toes are much more evident than in the other species.

Bullfrog *(Rana catesbeiana)*   PLATE 2

Adult Size. 3.5″ to 6.5″

Description. This frog grows to a greater size than any other species in the United States. It is usually very stocky, but smaller specimens are more slender. There is a ridge of skin running from behind each eye down alongside the eardrum (a large disc located just behind the

Bullfrog

Green frog

Northern leopard frog

Southern leopard frog

Pickerel frog

Mink frog

Wood frog

eye) to the top of the foreleg. The upper parts are olive or brownish yellow except for the head, which is usually a fairly bright green. Vague dark bars are usually visible on the hind legs. The underside is whitish with darker mottling here and there, and males have a yellowish throat.

Male's Call. The succession of deep, sonorous notes is usually described as sounding like "jug-er-rum." Whatever the descriptive phrase used, the depth and carrying quality of the call make it unmistakable. The number of notes uttered in each call varies from one to four.

Breeding. The bullfrog is a late breeder in the Northeast, usually from quite late in May through July. The female lays a tremendous number of eggs (up to 20,000) in a large, flat layer at the surface. This film of jelly-surrounded eggs may be more than two feet long and equally wide. Eggs hatch in a little less than a week to almost three weeks. The tadpoles grow to a length of more than four inches and do not transform into frogs until after their second winter.

Habitat. This aquatic species is seldom seen far from the shallows of the swamps, bogs, ponds, and lakes that it frequents. It seems to prefer bodies of water that are surrounded by trees, and where there are tree stumps and fallen trees lying in the water.

Comments. The record size for the bullfrog is usually given as 8", but one summer I caught a monster measuring 8.75" from a water hole in southeastern Arizona, where the species has been introduced. This species is very cannibalistic, and large specimens should not be kept together or with smaller frogs. I once had two large bullfrogs of equal size in the same container. One morning I found them both dead. The hind legs of one of them were protruding from the mouth of the other one, where it had evidently tried to swallow it and choked in the attempt.

Range in the Northeast. Southern Maine and throughout the rest of the Northeast (In some areas they seem to be becoming less common.)

Similar Northeastern Species. The green frog has ridges of skin extending down each side of the back.

Green Frog *(Rana clamitans melanota)*   PLATE 2

Adult Size. 2" to 3.75"

Description. The green frog is a fairly robust animal, with its hind toes webbed almost to their tips, except for the longest (fourth) toe, where the webbing reaches only to the second joint. A ridge of skin begins behind the eye and curves around the hind end of the large, rounded eardrum (located just behind the eye) almost to the shoulder. From the top of the eardrum, a branch of this ridge of skin extends down the side of the back. (This is called a dorsolateral fold.) The color is quite variable, but is usually greenish or brownish on the upper parts, with a green snout. There are dark brown spots and blotches on the back and sides, and dark brown bars are often present on the hind legs. The underside is white, usually with dark spots on the legs and along the sides of the belly. Males have a bright yellow throat.

Male's Call. A low-pitched, gulping note very much like the sound produced by plucking the string of a banjo indicates the presence of a green frog. There is usually a single note, but often the frog may repeat the note rapidly several times; each successive note is less loud, so that the call seems to fade.

Breeding. In the Northeast, breeding takes place from May to mid-August. Between 1,000 and 4,000 eggs are laid in a thin layer at the surface of the water. The tadpoles become very large, well over two inches in length, and do not transform into frogs until the spring or summer of the following year.

Habitat. This frog is found in just about every location where there is fresh water: around the edges of lakes, ponds, marshes, swamps, springs, temporary pools, along the banks of streams and ditches, and even in and around concrete bird baths built at ground level.

Comments. The green frog often squats in the shallows, but also enjoys sitting on the shoreline within easy leaping distance of the water. If alarmed, it will frequently utter a loud squawk as it jumps into the water. Males seem to establish and defend a territory. There is a large boulder at the edge of my pond that for two or three years has had a large green frog living near its base. During the day I have several times crept up on the inland side of this boulder and imitated a green frog's call. On each occasion that the frog has been present there has been an immediate reply, and upon my continually repeating the call the frog has moved around the side of the boulder almost to my feet. After imitating the call I once rustled a stick among the grass and the frog at once leaped on it.

Range in the Northeast. Common throughout the Northeast

Similar Northeastern Species. The bullfrog has very similar coloring, and a small or medium-sized bullfrog may easily be confused with the green frog. Bullfrogs, however, while having a ridge of skin around the eardrum, do not have the ridge running down each side of the back. This is an easy way of separating the two species. The mink frog is almost impossible to identify from the green frog except by its call. The toes of its hind feet are more fully webbed than those of the green frog, with the webbing extending to the last joint of the longest (fourth) toe.

Adult Size. 1.75" to 3"

Description. Pickerel frogs are slender bodied and smooth skinned, with a ridge of skin extending down the back from behind each eye. They are usually brownish. Between the dorsolateral folds are two rows of large, more or less square dark spots (sometimes with an extra spot or two tucked in between the two rows). There are more of these spots on the sides, bars on the legs, and a dark line from the eye to the nostril. The underside is white, except for the lower surface of the legs, which are bright yellow or orange.

Male's Call. The call is a single, grating low-pitched note lasting for a second or two, very much like a human snore. This sound does not carry very far. It may be heard by day as well as after dark, and is often delivered from underwater.

Breeding. Pickerel frogs breed from April to mid-May. The eggs are brownish and/or yellowish. There are 2,000 to 3,000 to a mass, and the masses are attached to underwater plant stems or twigs. Each mass is up to four inches in diameter, and several frogs may deposit their eggs in one small section of the pond. The eggs usually hatch in four to six days, and the tadpoles transform into frogs in about ten weeks.

Habitat. Although it is usually found in areas where the water is cold, this species has been breeding in and living around the edges of my pond for a number of years, even though the water temperature there is a good deal higher than in nearby streams. It occurs very commonly in sphagnum bogs and along rocky gorges, but like the leopard frog is often found well away from water after its breeding season.

Comments. This frog secretes a substance through its skin that is distasteful to animals that take it into their mouths. Do not put pickerel frogs into the same container as other frogs, for the skin secretion is strong enough to permeate the water and kill them.

Range in the Northeast. Throughout the Northeast, but missing from some areas

Similar Northeastern Species. The leopard frog has rounded spots, no dark line from eye to nostril, and no orange on the underside of the legs.

Adult Size. 2″ to 3.75″

Description. The body of this frog is relatively slender, with a prominent ridge of skin running back down the body from behind each eye. The skin is very smooth in texture. The upper parts are usually grass green, but sometimes brownish. Between the dorsolateral ridges are either two or three rows of large, irregularly spaced, rounded dark spots with light borders. There are more of these spots on the sides and legs, and a creamy line running along the upper jaw and over the top of the foreleg. The underside is whitish.

Male's Call. The male gives vent to a harsh, grating, low-pitched note lasting about three seconds, followed by several short, gutteral grunts. He usually calls while floating, but may sometimes call from underwater.

Breeding. This species breeds from April to mid-May. The eggs are laid in rounded masses that are flattened on the top and bottom and that are either attached to underwater objects or merely placed on the bottom of the pond. There are often communal laying sites where several frogs deposit their eggs in one spot. Each mass contains up to 2,000 or 3,000 eggs. The eggs usually hatch in less than one week, and the tadpoles transform into frogs in eight to eleven weeks, normally emerging from the water during July.

Habitat. Frogs of this species are found along streambanks (especially in meadow areas) and in ponds and marshy places. Once the breeding season is over they may turn up quite some way from water, and I have many times seen them amid the long grass in the middle of fields.

Comments. Leopard frogs are familiar to most biology students. They are commonly used for laboratory experiments, dissection exercises, etc.

Range in the Northeast. Common throughout the Northeast, but missing from some localities (including the Hudson Valley, parts of western New York, Long Island, and all but the northern parts of Connecticut)

Similar Northeastern Species. The pickerel frog has two rows of large, more or less square or rectangular spots down the center of the back, and some bright orange coloring on the underside of the legs. (This coloring is not visible when the frog is sitting.) It is almost always browner than the northern leopard frog.
　　The southern leopard frog has a more pointed head and snout,

and almost always has a whitish or golden spot in the center of the tympanum.

Mink Frog *(Rana septentrionalis)*   PLATE 2

Adult Size. 2″ to 2.75″

Description. Mink frogs are rather stout bodied, with slightly rough skins. The webbing on their hind feet extends to, but not beyond, the last joint of the longest (fourth) toe. They may or may not have dorsolateral ridges. Both color and pattern are quite variable. There is usually some green on the head, particularly on the lower jaw. The back and sides are brownish, with very dark mottling on the back outlined by yellowish lines. The legs are usually heavily spotted or barred, and the underparts of the body are white except in males, which always have a yellow throat and sometimes an entirely yellow underside.

Male's Call. The call is a loud, fairly low pitched "turtleac-turtleac-turtleac" repeated very rapidly up to ten times, and sometimes more. It may be heard by day as well as at night.

Breeding. This frog is a late breeder, the season lasting from mid-June to mid-August. The egg mass is often attached to the stems of water lily plants from one to two feet under water. This mass is about four inches in diameter and contains up to 2,000 eggs. Once hatched, the tadpoles transform into frogs about a year later.

Habitat. Mink frogs are aquatic. They are found in northern ponds, swamps, bogs, and lakes, but usually only where sphagnum moss and water lilies are abundant. I have often found them sitting on the floating leaves of water lilies in such areas.

Range in the Northeast. New York south to the Adirondacks, northern Vermont and New Hampshire, and all of Maine except the southern corner of the state

Similar Northeastern Species. Some green frogs are very similar in appearance to mink frogs, but the webbing on the longest toe on each hind foot of the green frog extends slightly beyond the second joint.

Adult Size. 1.5″ to 3″

Description. Wood frogs may be relatively slim or rather stocky. They have smooth skins and very prominent ridges of skin running back down the body from behind each eye. The coloring is buff, pink, or reddish above, with a large dark patch running back from behind each eye (rather like a mask) and a light line running along the upper jaw and back as far as the shoulder. There is also a dark line from the eye out along the snout, and dark bars on the legs. The underparts are shining white. During the breeding season the overall color may be so dark that the mask is not visible.

Male's Call. Wood frogs have low-pitched, clacking calls. Although a single frog usually will give no more than three or four of these notes in succession, a group calling together gives rise to a continuous sound rather like ducks quacking in the distance. The sound of such a chorus will carry for some little distance, but the individual call has little carrying power. Choruses occur during the day as well as at night. I once found a large number calling, under bright sun, from a small marshy area no larger than twenty feet square. Frogs were leaping around on all sides in the shallows, and calling while floating with arms and legs extended. In my pond I have many times seen the same frenzied activity after dark.

Breeding. These are very early spring breeders—in our area even earlier than spring peepers much of the time. I see them at the breeding ponds from mid-March to late April. Each female lays from 1,500 to 3,000 eggs in rounded masses about four inches in diameter. The egg masses are attached to underwater grasses or twigs, and this species often uses a communal laying site. On one occasion I found a sedge stem with five egg masses attached to it, one above the other. On the next weekend, when I again checked my pond, there were at least forty masses surrounding this particular sedge stem, and an area of about two feet in diameter was taken up by egg masses. There were so many that the tops of the upper masses protruded from the water. That night the pond froze, and with it the eggs above the water level. None of these eggs hatched, and it seems likely that many eggs may suffer a similar fate when there is freezing late in the spring. The eggs hatch in about two weeks, although if the weather remains cold, I have known them to take up to twenty-six days. The tadpoles transform into frogs in six weeks to almost three months.

Wood frog egg masses on a sedge stem

Habitat. During its short breeding season, this species is to be found in woodland ponds (or in ponds with woods nearby) and in almost any small woodland pool or marshy area. After breeding, the frogs move away from these sites, and may be found almost anywhere in the woods, where their coloring blends perfectly with that of dead leaves on the forest floor. I usually discover one or two in my vegetable garden during the course of the summer. Presumably they are seeking the dampness of the soil beneath the heavy hay mulch that I lay down between the rows of vegetables.

Range in the Northeast. Throughout the Northeast, mostly in wooded areas

Similar Northeastern Species. The dark patch behind each eye, coupled with the unpatterned back, make this frog extremely easy to identify. No other Northeastern species looks anything like it.

Other Northeastern Frogs

These are mostly small in size. Their ranges in the Northeast are very limited, although they may be locally common in the areas where they occur.

Northern Cricket Frog *(Acris c. crepitans)* PLATE 1

The adult size is 0.75″ to 1.25″. This is a small, blunt-headed frog. The top of the snout may be green, brown, red, gray, or almost white, and from behind each eye a stripe of the same color angles back to join and continue down the back as a single wide stripe, leaving a dark triangle between the eyes. There are four light lines angling toward the eye on each side of the lower jaw. The underside is whitish, with a darker throat. The call is a "click-click-click-click," rather like a cricket, and may be imitated by "stepping off" a few coins and then running another coin down the edges of them. This frog breeds from May to July. In the Northeast it is found in New York's Bear Mountain State Park, on Long Island, and south into New Jersey.

Western Chorus Frog *(Pseudacris t. triseriata)* PLATE 1

Adults are 0.75″ to 1.5″. There are three wide brown stripes down the paler back and another down each side. There is also a light line along the edge of the upper jaw. The underside is whitish, sometimes with some spotting on the throat. Breeding takes place in April and May. The call is a "prrrrrrreeeeeeep," rising in pitch at the end and sounding rather like the noise produced by running one's fingers along the teeth of a pocket comb. It is found in extreme western New York.

New Jersey Chorus Frog *(Pseudacris triseriata kalmi)* PLATE 1

This frog is like the western chorus frog, but is a little stockier and has the three stripes down the back wider and better defined. It occurs on Staten Island (New York) and south into New Jersey. The call is similar to that of the western chorus frog.

Upland Chorus Frog *(Pseudacris triseriata feriarum)*

This, again, is very similar in appearance to the western chorus frog, but it has the three stripes down the back often broken into blotches. There is usually a triangular marking on top of the head, between the eyes. Upland chorus frogs are found in northern New Jersey and eastern Pennsylvania. The call is similar to that of the western chorus frog.

Southern Leopard Frog *(Rana utricularia)* PLATE 2

While very much like the northern leopard frog, this species has a more pointed head and snout. Also, it almost always has a whitish or golden spot in the center of the tympanum. Its Northeastern range consists of extreme southern New York (including Long Island), northern New Jersey, and eastern Pennsylvania. The call is similar to that of the northern leopard frog.

Salamanders

An Early Morning Encounter

In the northeastern United States spring is sometimes slow in arriving. There are frequent late cold snaps, and in some years a snowfall comes even as late as mid-May. But on this particular May morning, there is no sign of snow, only a nip in the air and a cloudless sky.

I get up early and head for the ridge at the farm's northeastern boundary. The rising sun promises more warmth soon, and already along the crest of the ridge its light is stroking the treetops. Amid the green haze of fresh new leaves I make out the small, moving shapes of warblers, busily seeking insects. They have been flying for most of the night, heading toward their northern nesting areas, and they have settled here to rest and feed in readiness for the next stage of their long journey. As they move and feed, their sibilant, buzzy songs sound from all sides. They seem to move randomly. They flutter momentarily to the lower foliage or dart back to trees they have not yet explored for food. But always their overall movement is slowly northward. The urge to reach their summer home is strong within them.

The sun's rays have not yet reached the slopes below the ridge, where the woods are still shadowy and wet with dew. There is little movement, little sound. A red squirrel runs along a branch over my head and disappears behind a tree trunk. From a small marsh at the foot of the hillside comes the song of a red-winged blackbird, but the cheerful notes are shrunken by distance, muffled by the trees.

A large shelf of shale can be seen among the trees on the hillside. A small clump of columbines grows near it. Their dainty scarlet and gold blossoms add a delicate touch of beauty to the gray stone. At the foot of the shelf, large chips of shale lie scattered, fallen from the parent rock. Many of these pieces were dislodged when plant roots forced themselves into cracks of the rock; others were pushed out by the constant freezing and thawing of water in the cracks.

Some of these pieces of stone have fallen so that they are not flat against the soil, and I gently tilt a few so that I can see beneath them. Small hollows are revealed. They are cool looking and deep shadowed, like tiny caves. Dampness clings to the rock ceilings, and the soil floors are soft and moist.

In one such cave I discover a small animal. It is about five inches long and looks somewhat like a lizard. But this is merely a surface similarity; there are several things about it that are not lizardlike. Lizards are reptiles, and therefore have scales. This creature has no scales. Northeastern lizards have claws and ear openings. This animal has neither of these. Nor is this the right kind of habitat for a lizard; it is too moist.

The animal I am seeing is a salamander, a close relative of frogs and toads. This particular salamander is handsomely marked. Its soft body is almost black in color, patterned above with large, round, primrose-yellow spots. Thus its name—the spotted salamander.

The salamander has long since journeyed to the breeding pond. Its eggs have hatched into larvae, and the larvae have wriggled through the jelly envelopes that surrounded them. They are now to be found swimming freely in the pond. Having completed its reproductive activity for the year, the salamander lies motionless under the rock. It will remain there throughout the day, cool and moist like its surroundings. When dusk comes it may venture forth in search of earthworms and other food, but it will probably not move very far. Now that its annual migration to the pond is over, it has reverted to its usual slow pace of life.

As a group, today's salamanders are unknown to many people. Most salamanders lead a hidden, sluggish sort of existence. Yet many species are quite common. Some are found in wooded environments quite far from water; others live alongside streams where there are quiet backwaters and where rocks and tree stumps provide cover for them. Some spend their entire lives in water; others never enter it. Some have gills; others have lungs. Some have neither gills nor lungs, they obtain the oxygen they need directly through parts of their integument. Salamanders, in fact, seem to be a group of animals in which exceptions are the rule. They defy blanket description.

I carefully lower the piece of shale over the salamander and go my way. As I walk I muse upon the creature I have just seen. I think about its ancestors. They lived some 300 million years ago in a world where no backboned animals save themselves walked the land. The fossil record of these early amphibians indicates that they developed from strange-looking, lobe-finned fishes. Gradually, in the grindingly slow process of evolution, a group of these fishes developed lungs and began to breathe air. Eventually some emerged from the ancient seas of those days, and after more changes were able to walk on short, stubby legs. For a relatively short period in geologic time these lowly creatures ruled the land. They had no competitors other than their own kind. Then, from some of them, other forms of life evolved. One of the first great advances took place when a group of amphibians gave rise to animals that could lay their eggs on land. These new creatures were therefore even more independent of the water. We know them as early reptiles, and from them in due course evolved mammals and birds.

So the salamander that I had watched lying so stolidly in its damp hollow beneath the rock was a descendent of a truly remarkable group of animals. These were the first vertebrates to conquer the land, and as such they are ancestral to us all.

General Information

Of the more than 270 species of salamanders living today, and making up the order Caudata, there are about 90 species found in the United States and Canada, of which 16 occur in the Northeast.

Most salamanders are shaped like lizards, but unlike lizards they do not have scales, ear openings, or claws, and their skins are almost always moist. A tail is present in all stages from larva to adult.

Except in one species (the hellbender), fertilization of the eggs is internal. The male deposits small, jellylike structures called spermatophores, which contain live sperm. The female later maneuvers herself over a spermatophore and uses the lips of her cloaca (the common opening for the genital and excretory systems) to transfer it into her body. The sperm will later fertilize the eggs.

Depending upon the species, eggs are laid either in water or in moist places on land. Each egg is surrounded by one or more jelly envelopes. Some species lay their eggs singly, some deposit small clusters, and some produce larger masses of up to 275 eggs. The hellbender deposits long, beadlike strings. In species that lay their eggs in water, the eggs hatch into larvae that look like miniature adults except for having external, feathery gills. In some of the species that lay their eggs on land, the larvae develop completely inside the eggs, and may emerge with no gills at all. It may take several years before the adult stage is reached.

The food eaten by salamanders varies considerably, depending upon size, the time of year, the location, and other factors. In general, it is made up of insects and insect larvae, spiders, worms, small aquatic life, etc. Because of the similarity of foods eaten by salamanders, there is no separate food section given under any of the accounts of individual species that follow.

These are secretive animals, remaining hidden under rocks, logs, fallen leaves, and the like for most of the time. They are active mostly in late evening and at night. Some species make annual migrations to ponds and other quiet waters for breeding and egg laying, but unlike male frogs and toads, which call loudly during their breeding season, all salamanders are completely voiceless. In the Northeast most species hibernate.

Although the mudpuppy and some other salamanders are frequently used in laboratory exercises for studies in comparative anatomy, man has made little use of the group as a whole. Nor do salamanders appear to be important food items for most animals, although they may be eaten by mammals such as skunks and opossums and by some snakes and birds.

Probably the most common story in folklore concerning salamanders is to the effect that they can live in fire. This belief almost certainly originates from the fact that an occasional log will have a salamander hiding under its bark. If this log is put on the fire the heat causes the salamander to leave in a hurry, giving the appearance of having been living among the flames. Another common belief in some areas is that salamanders have "stingers" in their tails. This is quite untrue. None of these animals presents any danger to man.

Keeping Salamanders in Captivity

Most land-living salamanders can be maintained rather easily in captivity. A five-gallon aquarium tank can be a fine container. Make a lid from a sheet of heavy plastic with air holes punched with an icepick or similar sharp-pointed tool. Put some pebbles in the bottom of the container and cover them with a few inches of soil. Moss, small ferns, or other plants can help make an attractive terrarium. Be sure to place several flat stones or pieces of bark so that your salamanders can get under them and away from the light. Water requirements may be met by merely sinking a saucer or other removable shallow dish into the soil and half-filling it with water. Change this water two or three times each week.

Salamanders usually do well on a diet of small earthworms, caterpillars, or any other small, soft-bodied animal life that is available. Mealworms make a good stand-by item, but they should not be given as a steady diet due to their tough, sometimes indigestible skins. Do not merely drop food into the container; it will probably not be found by the salamanders. Hold the food in a pair of forceps and move it slowly back and forth in front of the animal. Touching it to the corner of the mouth will often stimulate a salamander to grab for it. Once it is used to feeding in captivity, try small pieces of raw fish or raw hamburger meat. Try to convince the salamander that its food is alive.

Feeding a salamander

For newts, which are aquatic as adults, use the same aquarium set-up you would have for fishes. Put in some large rocks that protrude from the water. Feed the newts by dropping small pieces of raw hamburger into the water. Newts are usually good feeders, but you should make sure each piece of meat is eaten before dropping in more, otherwise the water becomes fouled.

If you use city tap water for your aquarium, be sure you always let it stand in an open container overnight. This allows the escape of any chlorine that may have been added to the water.

Adult Size. 11.5″ to 17″

Description. This aquatic salamander has a heavy body with a groove running down the center of the back and three pairs of large, red, bushy, external gills. Its tail is very flattened vertically, with a fin-like crest. There are four toes on each foot. The color varies, but is usually gray-brown with some blurred, rounded, blue-black spots over the back and sides. The belly is light gray, often with dark spots scattered over it. An indistinct dark stripe extends from the snout through the eye and back along the head. The tail is often tinted with light red along the margins.

Breeding. From late September through November the male mudpuppy deposits spermatophores on the bottom of whatever aquatic locality it is inhabiting. The female waits until late spring before laying the fertilized eggs. The eggs, each surrounded by three envelopes of jelly, are laid singly and attached to the underside of submerged rocks or logs. One female lays from 20 to more than 175 eggs, each measuring about 0.25″ in diameter. She then remains at the site and guards the eggs. The incubation period is from five to nine weeks, at the end of which time the larvae, measuring about 0.75″, wriggle free from the jelly. They are colored differently from the adults, usually with a dark stripe from the snout to the tip of the tail. On each side of this stripe is a narrower, yellowish stripe. Mudpuppies are able to breed at about five years.

Habitat. The mudpuppy is strictly aquatic. It occupies a variety of water sites: clear lakes, ponds, and streams, but also weedy and muddy waters in ditches and the like.

Range in the Northeast. Vermont, western New Hampshire, Massachusetts, Connecticut, and New York except the south and south-central sections

Similar Northeastern Species. The hellbender is also large and aquatic, but it has a broad, flat head and a wrinkled fold of skin along each side of the body. Adult hellbenders have no gills, and there are five toes on each hind foot instead of four as in the mudpuppy.

Adult Size. 2.75″ to 3.75″

Description. The body of this salamander is quite slender. The head is fairly narrow, with a pointed snout, and the tail is very flattened vertically. In the adult (aquatic) stage there is usually dark green on the back and upper sides, but there may be brown instead. The underside is yellow. Small black dots occur over both upper and lower surfaces, and there are two irregular rows of larger red spots down the back. These spots have black borders and vary in both number and size from one animal to another. The male's tail becomes larger and rather finlike in the spring.

    The immature (terrestrial) stage is called a "red eft." It is bright orange-red on the upper surface, usually lighter on the underside. As in the adults there are red spots with black borders on the back, and sometimes one or two spots on the sides. The brightest efts are usually found in damp wooded areas at higher elevations. The skin is much rougher and drier than in the adults. The size of the red eft ranges from 1.25″ to 3.50″.

Breeding. Breeding occurs in the spring. The eggs are deposited one at a time, usually on leaves or stems of water plants. Each egg measures only about an eighth of an inch in diameter, including its surrounding jelly envelopes. After hatching, the larva remains in the water until late summer or early fall, and then moves out onto the land. This begins the "red eft" stage. It may last for as long as three years before the transformation into an adult is completed. The animal then moves back into the water and spends the rest of its life there.

Habitat. The adults, being aquatic, are found in ponds, marshy areas, and the like. They may also live in streams, but only where the water is very slow moving. The terrestrial efts are most commonly found in moist, wooded areas. They can often be discovered by turning over rotting logs. (Remember to return the log to its original position; it is probably the home of a good many other forms of animal life.)

Comments. I have several times seen adult newts in winter, swimming slowly beneath the ice on my pond. There are also reports of active newts being seen in winter in shallow field wells. These are exceptional sightings, however, and probably the vast majority of newts are true hibernators.

Range in the Northeast. Throughout the Northeast wherever the habitat is suitable

Similar Northeastern Species. None.

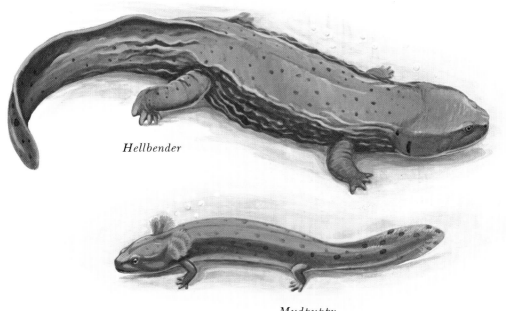

Hellbender

Mudpuppy

Jefferson's salamander

Spotted salamander

Marbled salamander

Northern red salamander

Eastern tiger salamander

Adult Size. 4.5″ to 7″, and occasionally larger

Description. In spite of its size, this salamander has a fairly slender body. The tail is oval at the base but compressed toward the tip. The toes are long and slender. The upper parts are almost black to dark brown; the underside is lighter. There are large numbers of small, light blue speckles, mostly along the sides, but also on the tail, legs, and throat. These blue flecks are often difficult to see, but are almost always present.

Breeding. The Jefferson's salamander breeds in early spring, often in the same ponds as spotted salamanders, and at the same time. Although one female may deposit up to 200 eggs, she lays them in several masses in different locations. Each mass contains from seven or eight to about 40 eggs, although I once found a mass with 63 eggs. The masses are almost cylindrical in shape and are attached to twigs or stems below the surface of the water. Each egg has several envelopes of jelly around it and the entire mass is surrounded by another thick layer of jelly. The eggs hatch in four to six weeks, and sexual maturity is reached at two years.

Habitat. Jefferson's salamander lives in wet areas within wooded sections, but is absent from evergreen forests. I have also found this species beneath logs in fairly open valleys that have been cleared of trees by beavers.

Comments. Several times, I have seen adults swimming beneath thin ice at the edge of breeding ponds. During this early part of the breeding season they are very easy to catch, probably because the coldness of the water slows them down.

Range in the Northeast. Present throughout the Northeast, but in rather scattered pockets

Similar Northeastern Species. The blue-spotted salamander has spots, rather then flecks, and is usually darker overall. It also has shorter toes than Jefferson's salamander. Unfortunately, this species sometimes breeds with the Jefferson's salamander, resulting in hybrids that are intermediate in color and markings between the two species.

The slimy salamander is less robust and has large white flecks.

Adult Size. 5.5″ to 7.5″

Description. The spotted salamander has a stocky body and a broad, rounded head with a wide snout. The entire upper ground color is black, often with a deep bluish tinge or sometimes with a dark brown tinge. Running from the head to the tip of the tail are two irregular rows of large round spots. These are usually bright primrose yellow, but may be darker, sometimes distinctly orange. Other spots may be scattered along the sides, and even on the legs. They usually number about twenty-five. The underside is a uniform slate color, quite a bit lighter than the upper surface.

Breeding. This is an early breeder, moving to the breeding sites from the end of March on into early April. I have found this species in ponds when there is still a thin film of ice covering part of the water. Other sites selected are marshes, bogs, and even temporary wet areas. For several years I found spotted salamander eggs in a low area where rainwater collected alongside my driveway. As with other species, the males deposit spermatophores on waterlogged twigs and leaves, etc., and the females later pick them up. The eggs are laid in masses measuring about three inches in diameter, each mass containing from 12 to well over 100 eggs. Each egg is surrounded by jelly, and in addition there is a jelly envelope around the entire egg mass. The egg masses are attached to sedges or to the stems of water plants or other underwater objects. Depending upon the water temperature, the eggs hatch in four to seven weeks.

Habitat. Aside from the annual period when adults may be found in the water at the breeding sites, they may sometimes be found beneath rotting logs, rocks, etc., in damp woods. They seem to be missing from evergreen forests, but almost any other wooded area is suitable. I once found several spotted salamanders under rocks on a relatively open hillside. This was in late April, however, and these salamanders were probably in the process of moving away from a nearby breeding pond.

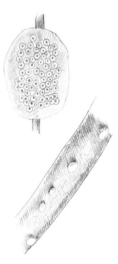

Spotted salamander spermatophore and egg mass

Comments. Both sexes sometimes participate in a form of aquatic nuptial dance. I have never been fortunate enough to see this, but a friend of mine witnessed such a performance in northern New York on 20 March 1976. He said that approximately sixty large adults took part. When first seen, they were scattered over the bottom of a shallow pond at a point where a small stream flowed into it. They then gathered into a compact group, and for about ten minutes swam rapidly together, rubbing against each other and swirling around until the water seemed to be boiling. The males had apparently already deposited their spermatophores, for a thick mat of dead leaves on the bottom of the pond was liberally sprinkled with them. At the end of this period of frenzied activity the salamanders disappeared among the dead leaves on the bottom.

After the first warm rain in early spring I have seen scores of spotted salamanders crossing the highway en route to their breeding sites. Where a heavily traveled road runs alongside such a site the pavement will often be dotted with squashed, run-over salamanders during the migration period.

I have found as many as thirteen egg masses in one location in a pond, actually touching each other. While a single female may lay several egg masses, this many makes me wonder if there are occasional cases where several females may be using the same spot as a communal egg-laying site.

Range in the Northeast. Throughout the Northeast in suitable habitats, but not commonly seen except during the breeding season

Similar Northeastern Species. The tiger salamander has a yellowish underside, and the yellow markings on its upper surface are very irregularly shaped. They may form bars on the tail and along the sides.

The slimy salamander has a more slender body than the spotted salamander. Its spots are white, more widely scattered, and much smaller.

Marbled Salamander *(Ambystoma opacum)*   PLATE 3

Adult Size. About 4″ but occasionally larger

Description. This is a stocky-bodied salamander with a relatively short tail, wider above than beneath. The ground color of both upper and lower surfaces is black, sometimes with a brown sheen under the head and on the legs. Running across the head, back, and tail is a series of white bars, which may be complete or broken and which may unite along each side of the animal to form a broken white stripe. These markings are bright white in males, grayer in females.

Breeding. Although related to the spotted salamander, which breeds in the spring, this species breeds in the fall, and the eggs are laid on land rather than in water. The site chosen is a small hollow that will fill with water during a heavy rain. If there is not enough rain in the fall to inundate the eggs, they will not hatch until the rains of the following spring. The eggs are laid singly, but all in one group. The female remains with the eggs.

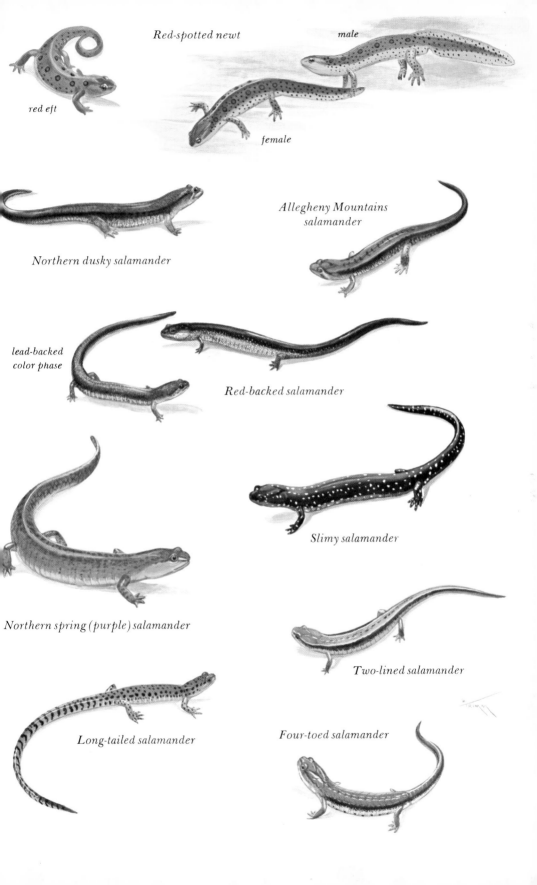

Red-spotted newt

male

red eft

female

Northern dusky salamander

*Allegheny Mountains
salamander*

*lead-backed
color phase*

Red-backed salamander

Slimy salamander

Northern spring (purple) salamander

Two-lined salamander

Long-tailed salamander

Four-toed salamander

Habitat. This species seems to live in almost any type of habitat. I have found it under rocks on open hillsides, beneath the debris of old planks and bricks in house foundations, in gravel pits, and under logs and rocks in quite moist woods.

Range in the Northeast. The marbled salamander occurs locally in southern New Hampshire, central Massachusetts, Rhode Island, Connecticut, southern New York (including Long Island), New Jersey, and eastern Pennsylvania

Similar Northeastern Species. Because of its chunky, startlingly black and white patterned body, this salamander is easy to identify. There are no other Northeastern species with which it can be confused.

Northern Dusky Salamander *(Desmognathus f. fuscus)*   PLATE 4

Adult Size. 2.5″ to 4.25″ (average 3.5″)

Description. Dusky salamanders are slender, but with a touch of robustness. The tail becomes progressively more vertically flattened and thin toward the tip, and is not quite half the entire length of the animal. The color and pattern are very variable indeed. The upper parts are colored almost any shade from light brown through reddish to black. Often there is a light band, bordered raggedly with black, running along the back and tail, but in darker-colored animals this cannot be seen at all. The underparts are flesh colored or pale yellow with grayish mottling. There is a pale line running diagonally downward from the rear of the eye to the rear of the mouth opening.

Breeding. The eggs are laid in summer beneath rocks, logs, or pieces of bark. They are deposited in clusters of about sixteen and are guarded by the female, who usually curls herself around them. Each egg is surrounded by three jelly envelopes. When the larvae hatch they have grayish backs with a light stripe running centrally the length of the body and tail. On each side of this stripe is a row of rounded light spots.

Habitat. Although sometimes found beneath stones in shallow water, this is essentially a terrestrial species. However, it always lives close to water, mostly along the edges of streams in wooded or rocky areas, but also around the edges of springs, along the shoreline at the foot of small waterfalls, and where water constantly trickles from seepage. It is almost always found beneath stones.

Comments. For a salamander, this is a very fast-moving animal when one is trying to catch it, and will wriggle free time and time again if not held securely.

Range in the Northeast. Missing from northwestern New York and northern Maine, but otherwise found throughout the Northeast

Similar Northeastern Species. The "lead-backed phase" of the red-backed salamander is more slender and has much black and white speckling on its underside.

The two-lined salamander and the Allegheny Mountains salamander both have dorsal bands extending down the length of the body and tail, but both have well-defined dark borders to the band and (usually) well-defined dark markings within the band.

Red-backed Salamander *(Plethodon c. cinereus)* PLATE 4

Adult Size. 2.25″ to 3.5″

Description. The body is long and slender, and the tail is round along its entire length. Although its color remains the same throughout its life, this salamander is found in two different color phases. The typically colored animal has a wide stripe running from the head down the center of the back and out along the tail. This stripe, which covers almost the entire back, may be any shade of pink or red, or even yellow or gray. There are often tiny black dots within the stripe. The sides of the body are black or very dark gray. The underside is a lighter gray, with many black and white flecks.

The other color phase is called the "lead-backed" phase. The broad dorsal stripe is missing, so that the animal's back is uniformly black or very dark gray. The sides are usually a lighter gray with black flecks. Individuals may be intermediate between these two color phases, where the dorsal stripe may be greatly broken. The two color phases are usually present in about equal numbers, but sometimes one or the other may be more common in a given area.

Red-backed salamander egg mass and young

Breeding. The male deposits spermatophores under rotting logs and rocks in the fall, and they are picked up by the female in the same season. The eggs are not laid until the following June or July. During this period, small groups of eggs are laid in little cavities under rocks and logs. They are attached to the roof of such cavities by a little stalk, and look somewhat like miniature bunches of whitish grapes. There are usually about eight eggs per bunch, but this number may vary from three or four to twelve or thirteen. The female remains with the eggs during the one to two months it takes for the eggs to develop into larvae. There is no aquatic larval stage. The young are small facsimiles of their parents except for having large external gills. These gills disappear after a few days. The young are able to breed at two years.

Habitat. This species is entirely terrestrial, and is common in wooded areas. As with most land-living salamanders, the best way to find one is to turn over rocks and rotting logs. Although usually occurring in moist woodlands this species can sometimes be found in other areas. I once discovered several red-backed salamanders beneath a pile of old planks in very dry, sandy soil.

Range in the Northeast. Found throughout the Northeast; probably the most common and widespread species in the area

Similar Northeastern Species. The dusky salamander may be colored somewhat like the lead-backed phase, but has a lighter underside with little or no speckling. The body and tail of the dusky salamander are not as slender. It lives along the edges of streams in much moister habitats than the red-backed salamander.

The Allegheny Mountains salamander has a broad, light brown stripe down the center of the back, but there is a row of dark markings—often chevron-shaped—running down the center of the stripe. The underside is flesh colored with few, if any, markings. In the Northeast it is limited to central, southern, and western New York west of the Hudson River.

Slimy Salamander *(Plethodon g. glutinosus)*   PLATE 4

Adult Size. 4.5″ to 6.75″

Description. Although it is quite a large salamander, this species has a slender body. The tail is entirely rounded, and is more than half of the animal's total length. The upper parts are shiny black or blue-black with large numbers of white flecks and speckles. The underside is slate gray; the throat a little lighter. In young animals the speckling may be yellow rather than white.

Breeding. It is thought that in the North this would occur in early spring. The eggs, numbering from seven to eleven, are laid on land. The entire group is suspended by a stalk from the underside of a log, and is guarded by the female. Although it has not been proven, there is a theory to the effect that northern slimy salamanders may lay their eggs below ground, before emerging from hibernation. Most members of this species breed when three years old, but some males may be mature at two years.

Habitat. The slimy salamander seems to prefer moist, wooded areas and hillsides. I have also found this species beneath piles of cut logs at the roadside and under pieces of shale half-embedded in wet banks.

Comments. This is an *extremely* slimy animal! It is very difficult to remove the slime from one's hands. I have used gasoline with some success, but when wishing to capture a specimen I now make a point of using a little stick to push the animal into a container.

Range in the Northeast. Pennsylvania, western New York, and the Hudson Valley south into northern New Jersey, but not on Long Island

Similar Northeastern Species. The spotted salamander is much more stocky and has large, rounded, yellow spots.

Northern Red Salamander *(Pseudotriton r. ruber)*   PLATE 3

Adult Size. 4.5″ to 6″; occasionally larger

Description. This is a stocky-bodied creature, with relatively short, thick legs. Its tail is noticeably shorter than its body. The entire upper surface is a shining, bright red or orange-red, dotted with many small black spots. The underside is lighter, almost flesh colored, with no spots. The eye is yellow. Older individuals are much darker, with a purplish sheen over the back, and with more numerous spots that merge into each other. Their undersides become more reddish, with small, dark brown spots.

Breeding. The eggs are laid in groups in the fall, and each group is attached by a short stalk to the underside of a stone beneath the surface in springs or other areas where there is clear water. Each egg has several jelly envelopes around it. The larvae normally lose their gills and transform into adults when they are about three inches long.

Habitat. This salamander apparently likes cold water, for it is found mostly in springs and in shaded streams. While it also lives in streams in more open areas, such streams have to be clear, without a muddy bottom. From June through August, adults may be found on land under rocks, stones, and logs.

Comments. I have found this species crossing roads during warm spring rains, usually in early April. Often they are crossing at the same time as spotted salamanders, and I have seen good numbers of both species.

Range in the Northeast. From southeastern and extreme south-western New York south into Pennsylvania and New Jersey; absent from Long Island

Similar Northeastern Species. The spring (purple) salamander is similar, but it is usually more yellow, with muddy, ill-defined dark mottling on the back and a very dark purple sheen. The sides are distinctly lighter than the back, and the tail is much more flattened toward the tip.

Two-lined Salamander *(Eurycea b. bislineata)*   PLATE 4

Adult Size. 2.5″ to 3.75″

Description. The body is slender with the snout short and blunt. The tail is compressed, sometimes with a keel. The ground color above is basically yellow, but may be orange-yellow or brown or even greenish yellow. A broad light stripe runs from the snout to the tip of the tail. Within this stripe are many small black spots. These are often concentrated along the center of the stripe, forming a broken line. From the back of each eye a wide black stripe runs along the sides of the body and out along the tail. The underside of the animal is a fairly bright yellow.

Breeding. Breeding occurs in late winter and early spring. The eggs are laid from April through early summer. They number about thirty and are attached individually to the underside of a stone or other object in running water. Several females may deposit their eggs at the same site. Each egg is surrounded by two jelly envelopes.

Habitat. This species is usually found along the edges of streams, sometimes beneath flat stones in the shallows, but most often under stones on the banks. The streams where these salamanders live need not be slow flowing, for I know of one colony that lives alongside a very fast moving stream at a point where there are miniature rapids. Occasionally individuals will be found in areas well away from water.

Comments. This species moves very quickly both on land and in the water. It will often leap out into the water and swim away to evade capture.

Range in the Northeast. Throughout the Northeast in suitable locations

Similar Northeastern Species. Dusky salamanders are sometimes light brown, and may occupy the same type of habitat. The edges of their dorsal stripe are not well defined, and there is a light line from the rear of the eye to the rear of the mouth opening.

Other Northeastern Salamanders

Some of these may be found in certain localities in the Northeast, but they are not as common as those salamanders previously described.

Hellbender *(Cryptobranchus alleganiensis)* PLATE 3

The largest salamander in the United States, the hellbender can grow to about twelve to eighteen inches in length, and some females are known to reach twenty-seven inches. This is a strictly aquatic animal. It can be recognized by its broad, flat head, the wrinkled fold of skin along each side of the body, and its vertically flattened tail with a crest. It is found in the Susquehanna River system in New York and Pennsylvania and in the Allegheny River in New York.

Eastern Tiger Salamander *(Ambystoma t. tigrinum)* PLATE 3

The average adult length is about seven inches. The upper parts of its stocky body may be black or dark brown with olive or pale yellowish brown spots and blotches. These markings sometimes fuse to form bars, especially on the tail. Its belly is greenish yellow. The eastern tiger salamander may be found in the Albany region of New York, on Long Island, and in New Jersey.

Allegheny Mountains Salamander
(Desmognathus o. ochrophaeus) PLATE 4

The average length of adults is from 2.75" to 3.5". The body is slender, the tail round. A broad stripe, which may be light brown, yellowish, or reddish, runs down the center of the back. The sides of this stripe have straight edges. Within the stripe there is a row of small black spots, typically chevron shaped but not necessarily so. The sides of the animal are dark, but become lighter again toward the belly. It is found in New York west of the Hudson River (but is absent from northern New York) and in northern New Jersey and northeastern Pennsylvania.

Four-toed Salamander *(Hemidactylium scutatum)*   PLATE 4

This is a small species, two to three inches in length. There are four toes on both front and hind feet, and there is a distinct constriction of the body at the base of the tail. The upper parts are yellowish or reddish brown. The belly is startlingly white, with black spots scattered over it. This animal is usually found only where there is sphagnum moss, in scattered localities from southern Maine through the Northeast. It is not present in the more northern sections of New Hampshire, Vermont, or New York.

Northern Spring (Purple) Salamander
(Gyrinophilus p. porphyriticus)   PLATE 4

The adult length ranges from 5.5″ to 7.5″. The body is fairly stocky and the head is quite long. The tail is rounded underneath, but is vertically flat above, with a sharp edge near its tip. The upper parts are yellow-brown or purplish, with a dull red sheen and a tracery of poorly defined dark flecks and speckles. The underparts are flesh colored. A light line, with a darker line beside it, runs from the nostril to the front of the eye. This salamander is found in southern Maine and scattered throughout most of the rest of the Northeast except for the extreme northern sections of New York, Vermont, and New Hampshire. It is absent from Rhode Island and Long Island.

Long-tailed Salamander *(Eurycea longicauda)*   PLATE 4

In this species, which has a total adult length of four to six inches, the tail is almost twice as long as the body. The upper parts of the animal are light brownish yellow, with vertical dark bars along the sides of the tail. The underside is yellowish along the edges and darker down the center of the belly. The Northeastern range is made up of southern New York, northern New Jersey, and eastern Pennsylvania.

Blue-spotted Salamander *(Ambystoma laterale)*

The adult length is about five inches. This salamander may very easily be confused with Jefferson's salamander, with which it may hybridize. Its color is similar but a little darker, with spotting rather than bluish speckles. In addition, the body of the blue-spotted salamander is a little more stocky. Although scattered throughout its range, this salamander is reported from northeastern New Jersey, northern New York (and down the Hudson Valley to Long Island), eastern Massachusetts, western Vermont, and south-central Maine.

Reptiles

Victims of Misunderstanding

Stone wall and trees at Howbourne

HERE is a group of animals that many people view either with mixed feelings or with active dislike. Carolus Linnaeus, the Swedish naturalist who established the consistent use of our present system for naming plants and animals, was among those who had little use for reptiles—or amphibians either, for that matter. He described them as "foul and loathsome animals" and went on to refer to their "filthy skins," "offensive smell," and "squalid habitation." These are all quite ridiculous accusations. They just do not hold up.

Reptiles, in fact, are remarkably interesting creatures. Like amphibians they are ectothermic (cold-blooded), which means that their body temperatures vary depending upon their surroundings. Unlike amphibians, however, they do not have smooth skins; they have a covering of scales. Most reptiles lay eggs, but their eggs are vastly different from those of amphibians. More than 270 million years ago, amphibians gave rise to a group of animals that laid eggs containing a sac filled with liquid. The embryo developed in this liquid. In other words, a substitute had been found for the open water in which all eggs, until that time, had to be laid. Thus, these new animals—the reptiles—were freed from the necessity of returning to water to lay their eggs. The embryos of birds and mammals also develop in such a sac, even now. It is called the amnion. So this new type of egg was indeed a very great step forward in evolution. It allowed the embryo to grow within a watery, but private, little environment.

Most of today's reptiles are still egg-layers, although many reptile eggs have shells that are leathery, rather than hard and brittle like those of birds. Those reptiles that do not lay eggs bring forth live young, but without the placenta that is associated with live-bearing mammals.

There are many different species of reptiles in the world today—something like six thousand of them. But originally there were many more. During the Mesozoic era, which began some quarter of a billion years ago and ended about 65 million years ago, there were more major reptile groups than exist at present. In fact, this was a time when these animals were the dominant creatures on earth. They occupied every type of living space. Dinosaurs and other reptiles roamed the land, pterosaurs ruled the skies, and some reptile groups made their home in the ancient oceans.

Most of these early reptile groups had died out by the end of the Mesozoic era, but the reasons for their extinction are lost in the mists of time. Many theories have been propounded, including the possibility of climatic change and its effects in altering vegetation, competition with more efficient forms of life, and overspecialization. It may have been a combination of these and other reasons, but nobody really knows or probably ever will know.

There are five major groups of reptiles living today: turtles, crocodilians, snakes, lizards, and a peculiar-looking and very primitive animal living in New Zealand called the tuatara. Snakes and lizards are usually classified as one group, though there are those who feel it is more convenient to separate them.

In the Northeast we have representatives of the snakes, lizards, and turtles. Some of them are quite rare. Some are found only in small pockets, but are common enough where they do occur. Some are quite common and widespread. Most of those in our area are active during the daylight hours, and many are brightly colored. Why, then, are they so infrequently seen? There are several reasons. First, the reptiles of the Northeast all hibernate. Thus, we would not expect to see them during the winter months. Second, since they do not have to migrate to ponds for breeding and egg laying, there is no period when large numbers of them would come together in one place. (An exception would be the communal hibernating dens of snakes, but these are always well concealed.) Third, they are voiceless, and therefore do not draw attention to themselves. Finally, they are mostly quite secretive, tending to hide as much as possible.

An encounter with a reptile—any kind of reptile—in the Northeast is a rather unusual event. It should be relished to the full. Linnaeus notwithstanding, there is nothing foul, loathsome, filthy, smelly, or squalid about reptiles. He just had not bothered to learn anything about them.

Turtles

The Armored Wayfarer

The early October day is quiet, calm, and clear. From How-bourne, I can see the Catskills to the southwest, standing crisply against the skyline; in the foreground the countryside glows with fall colors. The reds and oranges and yellows of maples and ashes, oaks, and hickories are interspersed with patches of dark green where groves of white pine and hemlock stand. The summer just ended has bleached the grasses along the hedgerows and made them brittle. Tall stems bear the gaping shells of brown milkweed pods. Empty of seeds, the shells show their white linings, which shine as if varnished. Deep red seed stalks of dock and clumps of velvety-leafed mullen also project from the grass. On the open slopes, the gray dogwood leaves are now purple-red, and small bunches of waxy white berries dot the foliage. Bittersweet crawls over some of the dogwood, and the orange berries have burst open to reveal their red seeds.

The glaring heat of summer has passed, and already there have been nights of light frost. From weedpatch and brushland I hear the soft, seeping calls of white-throated sparrows. They are moving southward now, and sometimes one of them will attempt a song, but the song is half-hearted and off key. The breeding season is past, and there is no real urge to sing.

I walk across the hillside and into Beth Woods, on the western side of our farm. Here, at the edge of a small glade, I find a comfortable seat on the ground and lean back against a tree. The vegetable garden has yielded a good harvest and now I have the time to enjoy the languorous autumn beauty of the woods.

Among the multicolored carpet of fallen leaves near where

I am sitting there is a soft rustling. A box turtle appears, trudging slowly and ponderously across the open glade. It stops frequently, as if the weight of its shell were almost too heavy to bear. Sometimes, when it stops, it cranes its neck and peers dimly up at the branches over its head. Sometimes it just squats there on its stumpy legs and surveys the area on each side of it with great deliberation. Then it resumes its plodding path.

It clambers laboriously up over the edge of a small shelf of rock. When it is part way over the lip it almost topples backward, but its front legs scrabble furiously and eventually obtain a grip, and it is able to drag itself up and over the rock shelf. It rests again after this violent effort. A red and yellow maple leaf drifts down from above and lands just in front of it. At once its head disappears back into its shell, and the front of the shell closes. Minutes pass. With a series of jerks the shell opens slightly. The turtle's shining eyes appear, back in the shadows. Gradually the shell opens wider, and a little at a time the turtle's head slowly emerges. It scans the area suspiciously. Another leaf flutters to the ground nearby, and the turtle begins to draw in its head again. But this time it seems to realize there is nothing to fear. It glares malevolently in the direction of the leaf, pushes its head out completely and continues on its way.

A chipmunk pauses from investigating an old log. Its cheek pouches bulge with seeds it has been collecting. It sits and watches the armored reptile moving slowly by. The turtle reaches the far side of the glade and disappears into the undergrowth. The chipmunk returns to its search for winter food supplies.

Many of us are prone to look upon most species of turtles with some amusement. We admire their tenacity, but deplore their timidity. These hump-backed, shell-encased creatures inspire a vague affection in us, possibly because they appear to be so completely helpless in the universal fight for survival. Although we know that some species may reach a great age (though not as great as some people believe), we are not quite sure just how they achieve this with so little apparent effort. They seem to blunder through life more by accident than by any push on their part.

Maybe it is this very lack of aggressiveness in most species that is the key to their success. Certainly their protective shells have something to do with it. Whatever the reason, turtles are among the most successful of all animals. With little change in their general appearance they have persisted for something like 200 million years. They watched the rise of the dinosaurs and other great reptiles of the Mesozoic era, and they saw them fade into extinction. They beheld the first, primitive birds and the earliest mammals. Down through the ages they have lumbered their silent way, ubiquitous but unobtrusive, placid but persistent. Perhaps our amusement at these cumbersome creatures should be tempered with a little respect!

General Information

While there are about 250 species of turtles in the world, only 46 occur in the United States and Canada. Of these, there are 18 species in the Northeast. Turtles are classified as the order Testudinata.

All turtles have an external shell consisting of two distinct parts. The upper shell is called the carapace; the lower shell, which protects the underside, is the plastron. They are joined at each side by a "bridge." The male's plastron is usually slightly concave. This assists him in mounting the female during mating. She is usually larger than the male. The female's plastron is flat or even a little convex. In most species, both carapace and plastron are covered with large plates called scutes—or sometimes they are called shields. The skin over the rest of the body has scales.

Turtles have no teeth, but their horny jaws are hooked and sharp edged. They have claws on their fingers and toes, and they have tails, the latter being quite short in most species. All turtles are voiceless, but the wood turtle sometimes utters a soft whistling sound.

There is often confusion arising from the three terms "turtle," "tortoise," and "terrapin." Some terrestrial (land-living) members of this group of animals are often called tortoises, and some of the aquatic species are termed terrapins. "Turtle" is an all-embracing and quite correct name used for all of these animals. It would probably be simpler if the other two terms were dropped from usage altogether!

All Northeastern turtles hibernate, often in the mud on the bottoms of ponds. Turtles have a low rate of metabolism, and since in winter the cold water holds more oxygen, this is sufficient for their needs.

Mating usually takes place in the spring, but may continue on into the fall months. All species lay eggs, and one mating may fertilize eggs for several seasons, but the number of fertile eggs becomes less in each succeeding year unless there is another mating. In June or July the female uses her hind legs to dig a hole in soil, sand, or other soft material and there lays her eggs. The eggs are usually oval, but in some species they are almost round. In many species they have a distinctly rubbery shell. The number of eggs laid at one time varies from two or three to over thirty.

The eggs usually hatch during the summer, but if they were late in being laid the young may not emerge until the following spring. The young turtle has an "egg tooth" to help it escape from the shell, but it is thought that often the egg absorbs water and ruptures without much assistance from the hatchling. Newly hatched turtles have a soft shell, and are often colored very differently from adults. Once hatched they are on their own, for there is no care given to them by the adults. Turtles are able to breed when they are approximately five years old.

Because of their low metabolic rate, turtles can exist on a surprisingly small amount of food. Most species eat both plant and animal matter (fish, insects, and other small creatures), and some are scavengers, feeding on dead animals.

All turtles can swim, and some are particularly well adapted to an aquatic existence by having webbed hind feet. On the other hand, they are poor climbers, and are only occasionally able to drag themselves onto the lower limbs of bushes or trees. Turtles are active mostly in the daytime, and most of the aquatic species enjoy basking in the sun.

From an economic standpoint, these creatures are of some importance. Man has probably always used turtles and their eggs as food, and at least two Northeastern species (the diamondback terrapin and the snapping turtle) are eaten today. Large snapping turtles are thought to feed to some extent on game fishes, and even to drag ducklings beneath the water and eat them. While this predation may sometimes be true, such reports are probably exaggerated. Many stores sell turtles as pets. This has become so widespread that several of the more easily caught species are now protected by law in some states.

It is commonly believed that turtles have fantastically long lifespans. Although there are apparently authentic records of individuals of some species living for more than 150 years, this is quite exceptional. Captive turtles may live for 50 years or more, but most turtles in the wild probably do not live that long. Some people call all turtles "snapping turtles," believing that every species bites viciously. This is not so. It is true that some do bite readily, but of these the snapping turtle is but one species. A large snapping turtle is capable of inflicting a very nasty bite indeed, and it is best to pick it up by the rear of the carapace or by its tail. Hold it at arm's length to avoid being bitten on the body or legs.

Keeping Turtles in Captivity

The size of the container you select in which to keep turtles will, of course, depend upon the size of the turtles. For land turtles a large, fairly deep box such as an orange crate is quite suitable. Newspaper makes a perfectly adequate and easily changed floor covering. It is also a good idea to cut away the side of a smaller box and place it upside down on the floor of the container. This gives the turtles something beneath which to hide. Have a large, shallow pan of water always available to the turtles, and remember to change the water at least every other day. Feed a mixture of small pieces of apple, grapes, carrots, lettuce, tomatoes, or almost any other fruit or vegetables, plus small pieces of raw meat. Put all of this food into a shallow dish and let the turtles help themselves. Feed them on alternate days and about once a week rub a little cod-liver oil and bonemeal into the food. This provides the turtles with vitamin D and calcium. Turtles show much individual preference for certain foods, and only experience and experimenting with different foods will show you which items are favored by your specimens.

If you wish to keep aquatic turtles, use an aquarium tank about three-quarters full of water, with large rocks stacked in such a way that the turtles can rest on them, out of the water, if they wish. Change the water regularly. When feeding aquatic turtles, put them into a different container (no water needed) so that the water in the aquarium is not fouled by uneaten food. Feed mostly raw meat or fish, but also have lettuce or other greens available. When giving cod-liver oil and bone meal, follow the same procedure as with land turtles.

Adult Size. Carapace length about 3.5″ to 4″

Description. This small turtle has a relatively smooth carapace of a dull greenish or brownish color; the plates, or "scutes," do not overlap. The head has two yellowish stripes on each side, one above and one below the eye. The short tail ends in a blunt spine.

Breeding. The eggs are laid during June and early July, usually no more than five to a clutch. The site chosen varies considerably, and nests have been reported in muskrat houses, in sandy soil, in banks, and under logs; sometimes the eggs are merely deposited on top of the ground.

Habitat. Although seeming to prefer ponds with an abundance of aquatic plants, almost any fresh-water area may harbor this species. I have found musk turtles in slow-moving streams and even in muddy ditches where there was barely any water at all. They are usually to be seen on the bottom, and, although reported as being slow swimmers, can move quite rapidly when the need arises.

Comments. When picked up, turtles of this species usually hiss, open their mouths, and attempt to bite. Although small in size they invariably seem to have bad tempers!

Range in the Northeast. Southeastern Maine, southern Vermont, and New Hampshire west to western New York and south throughout the rest of New York, Massachusetts, Rhode Island, Connecticut, New Jersey, and eastern Pennsylvania

Similar Northeastern Species. The snapping turtle has a long tail and the rear of its carapace is deeply serrated. The painted turtles have red around the edge of the carapace, and the eastern mud turtle has small yellow markings on the head but no yellow head stripes.

Snapping Turtle *(Chelydra serpentina)*   PLATE 5

Adult Size. Carapace length up to 12″, sometimes larger; weight up to more than 30 lbs.

Description. There should be no confusion in identifying this turtle. The carapace is dark brownish, grayish, or blackish, and has very deep serrations at the hind end. For a turtle, the tail is very long—about the same length as the carapace—and has a jagged crest along its upper side. (It always puts me in mind of a dinosaur tail!) The head is large. On the underside, the plastron (lower shell) is small, light colored, and more or less cross shaped.

Female snapping turtle laying eggs

Breeding. In the Northeast, eggs are laid during June, usually in soft soil on the banks of a pond or marshy area. Although most eggs hatch in early fall, I once found three separate nests (in a long bank sloping down to an Adirondack beaver pond) in mid-July, where some of the young had hatched from each, while several unhatched eggs remained in the nests. Including the empty shells, the three nests contained eighteen, twenty-five, and twenty-seven eggs, respectively. At hatching the young have a carapace little more than one inch long. They are distinctly dull blue in color.

Habitat. Snapping turtles live in ponds, swamps, marshes, and slow-moving streams wherever there is some aquatic vegetation present. On one occasion I saw what first seemed to be a rock at the base of a series of rapids in a trout stream. It turned out to be a large snapping turtle braced against a rock, with the water foaming around it. As I watched, it lost or released its grip and was washed downstream for several yards before being able to struggle ashore. It then rested for a few minutes and then it staggered up the bank, crossed a highway, and disappeared into a swamp on the other side.

Comments. Individuals of this species almost always have ugly dispositions. The name of the animal is derived from the fact that, particularly when defending itself, it will shoot out its neck with great speed and endeavor to deliver a vicious bite. If successful in the attempt the sharp, hooked jaws clamp onto the victim and maintain a powerful grip. Obviously, a large specimen can inflict a bad wound, and it is as well to pick up this turtle by the tail, holding it well away from one's legs. Snapping turtles are often used for making turtle soup. It is tasty, but since most snappers contain large numbers of worms and other internal parasites the meat should be very well cooked!

Range in the Northeast. Throughout the Northeast

Similar Northeastern Species. None. This is the only Northeastern species that has a long tail and deep serrations on the hind end of its carapace.

Adult Size. Carapace length to about 4.5″

Description. The carapace is black or dark brown, normally marked with a sprinkling of small, rounded, primrose-yellow or orange spots. The number of spots varies tremendously, and occasionally there may be none present at all. If this is the case, look for spots on the head and neck. The plastron is dull yellow, orange, or dark colored, with some larger dark markings at the edges of the plates.

Breeding. Mating normally occurs in late April, and the eggs—from one to four in number—are deposited during the latter half of June. As with most turtles, the female digs a flask-shaped hole in which to lay her eggs. This hole is excavated in soft soil not far from water. The eggs hatch in early September.

Habitat. Almost any wet, fresh-water area is acceptable to this species, but it seems to prefer fairly shallow, muddy locations rather than lakes and other large bodies of water. I found one individual on a damp, wooded hillside, but this was the exception rather than the rule.

Comments. I have never known this turtle to bite. It enjoys sunning, although I have never seen large numbers basking at the same site as in the case of the painted turtles.

Range in the Northeast. Southwest Maine, southeast New Hampshire, and all of Massachusetts, Connecticut, Rhode Island, New Jersey, and Pennsylvania; missing from Vermont and northern New York

Similar Northeastern Species. None. The yellow spots on the carapace and/or head and neck identify this turtle at once.

Wood Turtle *(Clemmys insculpta)*   PLATE 5

Adult Size. Carapace length to about 7.5″

Description. The throat, nape of the neck, legs, and tail have much bright orange coloring. The brown carapace is rather flattened overall, but each separate scute is shallowly raised, more or less five-sided, and distinctly marked with concentric ridges and grooves. The smaller plates along the edge of the carapace are similarly marked.

Musk turtle

Snapping turtle

Spotted turtle

Bog turtle

Wood turtle

Northern
diamondback terrapin

Eastern box turtle

The large plastron is yellow, with six large, dark blotches down each edge (one on each of the plates).

Breeding. Mating takes place in late April or May, but fall matings have also been observed. Eggs are laid in mid-June, usually in holes excavated in sandy soil near water. There are normally from seven to twelve eggs, which hatch in late September or early October.

Habitat. As its name suggests, this turtle frequents wooded areas but it may also be found in fields and swamps. It swims well, but is much more likely to be seen plodding along through the woods or across the fields. I have several times found wood turtles run over on the highways and have removed live ones from such dangerous situations on a number of occasions.

Range in the Northeast. Throughout the Northeast, but missing from Long Island.

Similar Northeastern Species. The bog turtle is smaller and has a yellow or orange spot on the head. There is no orange on its legs or tail.

Eastern Box Turtle (*Terrapene carolina carolina*)   PLATE 5

Box turtle

Adult Size. Carapace length to about 6″

Description. The carapace is high and domed, with a keel running down its center. Each carapace plate is brown, marked with fairly bright, buff-yellow blotches and lines. There are yellowish markings on the throat. The large plastron is a plain golden yellow. It is hinged forward of its center, so that the turtle is able to close the front of the plastron against the underside of the carapace. Males often have bright red eyes, while the females' eyes are usually brown.

Breeding. From two to six eggs are laid during June or early July. The nest hole is usually in loose soil, and is about two inches deep. Both excavation and laying activities are carried out in late afternoon or early evening. The young emerge mostly in September.

Habitat. This is mainly a woodland turtle, although often it may be found in the moister parts of the woods. I once found one in an open field a good mile from the nearest woods, but this is probably quite exceptional, and every other individual I have found has been within well-wooded sections.

Comments. This species does not usually attempt to bite, although it may hiss if picked up. Some individuals are believed to live for more than 100 years. James Oliver, in his *Natural History of North American Amphibians and Reptiles* (1955), records one known to be 138 years old. Such data are probably the result of the practice of scratching initials and dates into the plastrons of this species, a practice very much frowned upon today. Certainly the box turtle seems to live longer than most other Northeastern turtles.

Range in the Northeast. New Jersey and eastern Pennsylvania, southeastern New York (including Long Island), Connecticut, Massachusetts, Rhode Island, and extreme southwestern Maine (It is possible that they also occur in extreme southern New Hampshire.)

Similar Northeastern Species. Blanding's turtle is larger, and has a bright yellow chin and throat.

Northern Diamondback Terrapin
(Malaclemys terrapin terrapin)

PLATE 5

Adult Size. Carapace length 4″ to 8.5″. Males are smaller than females

Description. Although the color varies considerably, there are two characteristics that make identification simple: (1) each plate of the carapace has a number of dark-colored, raised, vaguely hexagonal or circular lines forming a series of ridges and grooves, and (2) the head and neck, and usually the legs, are grayish with many darker spots and streaks.

Breeding. In the northern part of its range this species comes ashore to lay eggs from mid-June to mid-July, when hundreds may sometimes be seen at favored nesting sites. On 11 June 1961, I saw an estimated fifteen hundred to two thousand swimming just offshore in New Jersey. They were obviously preparing to come ashore, but I found none actually out of the water. A local fisherman told me that they usually lay eggs there during the first week of July, so apparently they were quite a bit earlier that year. The female digs a hole six inches deep in the sand, well above the high-water line, and lays up to twelve eggs. These hatch in late August, the newly hatched young having carapaces about 1.5″ long. Maturity is reached at about six years.

Habitat. This is strictly a salt- or brackish-water species. It is found along the coastline where there are salt marshes or where streams meet the sea. It will swim up such streams or rivers, and may be seen sunning itself on the mud at low tide, always within easy reach of the water.

Comments. At one time this turtle was greatly favored for its tasty flesh, and a price of about $7.00 would be paid for an average-sized specimen "on the hoof." As a result, the species was becoming rare by the end of the nineteenth century. Today it is protected throughout much of its range.

Range in the Northeast. Coastal areas from Cape Cod, Massachusetts, south into New Jersey; uncommon throughout most of its Northeastern range

Similar Northeastern Species. None. The ash-gray skin of the head and neck, together with its dark spots and streaks are unmistakable. This species is found only in salt water or brackish water.

Eastern Painted Turtle *(Chrysemys picta picta)* PLATE 6

Adult Size. Carapace length to about 6″

Description. The carapace is smooth and flattened, with no keel down its center. Its color is olive, brown, or blackish. The large scutes of the carapace have yellowish margins, and are in parallel rows down the back. (The rows do not alternate as in many other turtles.) There are yellow stripes on the head, red stripes on the neck and legs, and red and black markings on the bridge (where the upper and lower shells join together at the sides). The plastron is a uniform orange-yellow, and covers the entire underside of the animal. Like many turtles, this species, when handled, will often hiss and sometimes tries to bite.

Painted turtles sunning on a log

Breeding. During the mating season, which is usually in spring, the male develops very long claws on his front feet. He uses these to scratch the female's head and neck, thus presumably putting her into the mood for accepting his advances. The eggs are laid mostly during the last half of June and the first half of July. Sites selected for

Eastern painted turtle

Midland painted turtle

female

Red-bellied turtle

male

Eastern spiny soft-shelled turtle

Blanding's turtle

Eastern mud turtle

Map turtle

egg laying vary tremendously, but are usually in soft soil fairly close to water. Sometimes a nesting hole may be excavated in a lawn. As with most species, digging activity is usually in the evening, and the female frequently moistens the soil by urinating on it. The number of eggs laid varies from four to about twelve. Hatching occurs in late August or early September.

Habitat. Painted turtles may be found in almost any watery situation where there is plenty of aquatic plant life and where the water is not fast flowing. Although ponds, lakes, marshes, and swamps seem to be preferred, I have often found this species in slow-moving streams.

Comments. These turtles enjoy socializing and basking in the sun, and may often be seen in sizeable groups on rocks, logs, and exposed mud banks. Although apparently dreaming away the day when in such locations, they keep a sharp lookout and are very difficult to approach. I have often tried creeping up on such baskers, but almost invariably they slip into the water long before I get near them. It is thought that this sunning may help to rid the shell of algae, leeches, and other nuisances. Many times individuals may be seen swimming rapidly along a shoreline right at the surface, twisting and turning among the water plants and creating quite a disturbance in the shallows.

Range in the Northeast. Southern Maine, all of Massachusetts, Vermont, New Hampshire, Rhode Island, and Connecticut, and in eastern and central New York south into New Jersey and eastern Pennsylvania

Similar Northeastern Species. The midland painted turtle is very similar, and its range overlaps that of the eastern painted turtle to quite a large extent. Usually it can readily be separated by looking at the carapace scutes. In the midland painted turtle the rows of scutes alternate. In addition, the midland painted turtle has a large, oval, dark blotch on its plastron. In some parts of Long Island, the Hudson Valley, and northern New Jersey these two subspecies hybridize.

Midland Painted Turtle *(Chrysemys picta marginata)*   PLATE 6

Adult Size. Carapace length to about 5.5″

Description. This turtle resembles the eastern painted turtle except for two differences: (1) the large scutes of the carapace are *not* in parallel rows: the edges of the row of plates running down the center of the back alternate with the edges of the row to each side; and (2) the plastron usually has a large, more or less oval, dark blotch extending for most of its length.

Breeding. See the eastern painted turtle.

Habitat. See the eastern painted turtle.

Range in the Northeast. Missing from Maine, eastern Massachusetts, Rhode Island, and eastern Pennsylvania

Similar Northeastern Species. See the eastern painted turtle.

Other Northeastern Turtles

These turtles all have very limited ranges in the Northeast.

Eastern Mud Turtle *(Kinosternon subrubrum)* PLATE 6

The carapace is about four inches long and is brownish in color, with no markings. The head has small yellow spots, blotches, or streaks. There is a double hinge on the plastron. This turtle lives in marshy areas and in small, shallow ponds and muddy ditches. It occurs only on Long Island and the extreme southern tip of New York, and in the southwestern corner of Connecticut.

Bog Turtle *(Clemmys muhlenbergi)* PLATE 5

The carapace of this rare species is up to four inches in length. It is dark brown with a large, dull yellowish blotch in each large scute. The outstanding feature is a bright orange or yellow patch on the head (often a patch on each side of the head). The plastron is black with irregular yellow markings. The bog turtle is found in bogs and swamps in a few limited areas of New York, New Jersey, and Connecticut.

Map Turtle *(Graptemys geographica)* PLATE 6

This turtle may grow to a considerable size, with a carapace length of almost eleven inches. It is colored olive or brown with an intricate network of greenish yellow lines. There is a yellow spot behind the eye, and thin yellow lines run along the neck. The plastron is pale yellow and is unmarked. This species occurs in rivers, lakes, and other large watery areas. In the Northeast it is found in Lake George, Lake Champlain, the Great Lakes and St. Lawrence regions of New York, and extreme northwestern Vermont.

Red-bellied Turtle *(Pseudemys rubiventris)* PLATE 6

This large turtle has a fairly flattened carapace up to twelve inches long. The color is dark brown or black, mottled with red in males and with red lines in females. The plastron is yellowish or reddish, but always has red around the margin. There is a small population of this species reported to be living in ponds and streams in northeastern Massachusetts.

Blanding's Turtle *(Emydoidea blandingi)* PLATE 6

The carapace is up to eight inches in length and is black with a great many yellowish rounded or oblong small spots present. The chin and throat are bright yellow. The plastron is yellowish with dark blotches at the margin and is hinged near its center. While often seen walking on land, this turtle is found mostly in marshes, bogs, swamps, and ponds. In the Northeast it is found only from southeastern New Hampshire into northeastern Massachusetts.

Eastern Spiny Soft-shelled Turtle
(Trionyx spinifera spinifera) PLATE 6

Although the female's carapace may be sixteen inches long, that of the male is no more than nine inches. The carapace is grayish or brownish, tremendously flattened, leathery, and soft. There are no scutes present at all, but there are many small brown speckles and spots all over the back and feet. The snout is very long and is used as a "snorkel" for breathing without exposing the body. This turtle lives mostly in rivers, and in the Northeast it occurs only in northwestern and western New York.

Snakes

The Silent Predator

On hot summer days we look out from our hillside location near Lebanon Valley to where the ridges and hillsides shimmer with haze. They seem unreal and out of focus, a rolling mass of misty green swimming in the sunlight and fading in the distance to soft blue.

In one of the wooded valleys out toward the Hudson River there is a stream. Sometimes, to escape the heat, I make my way out through the fields and woods and wander for a while along the banks of this stream. It is one of my favorite walks. The stream winds among the trees and washes the rocks with gentle ripples, and it is cool and pleasant beneath the trees. At one point in the valley the stream emerges from the woods into a large clearing and flows quietly into a pond. Around the pond are many tree stumps, and out in its center is a beaver house. The beavers have long since abandoned the area, and the house they built has moss and grass growing from its conical roof. It pushes up tiredly, a small brown and green island in the shining expanse of water.

At the lower end of the pond the beavers constructed a long dam of logs, branches, rocks, and mud. Now the dam is broken. Floods have washed away a section near its center, and smooth water gushes through the gap. It splashes down into a hollow it has worn below the dam, and from this pool the stream is reborn and continues on its way.

Near one end of the dam there is a small boggy spot reaching out into the water. Here the silt gradually built up as it was washed down into the pond by the stream. When the dam was ruptured and the pond level dropped, the silt was exposed. Sedges and cattails crept out from the banks, and much of the rich mud is now covered with vegetation.

One July morning I was standing in this area, looking at the animal signs around me. The footprints of deer and foxes were everywhere. I could see where muskrats had been nibbling at the cattails. A line of tracks and overturned stones along the water's edge showed where a raccoon had been hunting for crayfish. Its tracks looked like little human hands with spread fingers.

Suddenly I noticed a leopard frog sitting near the shoreline. It was partially hidden by a small clump of sedges, within easy leaping distance of the water. When the soft breeze stirred the sedges the hot sunlight would touch the frog for a few moments. It sat placidly in the intermittent shade.

From among the tussocks, a water snake appeared. Its long, slender body slid easily over the ground like a silent runnel of brown, viscous liquid. It paused frequently, and with the rear of its body anchored to the ground thrust its head among the bases of the sedges. Its tongue fluttered from its mouth. The lidless eyes gleamed as the snake turned its head. It glided out from between two tussocks and there, a few inches in front of it, sat the frog.

The snake became motionless save for its flickering tongue. Its head was raised, and its eyes stared unwaveringly at the frog. Then, very slowly, the hind end of the snake's body began to move, to slide softly forward. As it did so the head lifted higher. There was a long moment as the snake readied itself. The frog sat motionless.

Suddenly the snake struck! Its head shot forward and its open mouth clamped onto the frog's head. The frog gave a loud squawk and leaped frantically from the ground, twisting and turning its body violently. Again and again it threw itself frenziedly into the air and thrashed against the sedges. But the snake had a firm grip. Its sharp, curved teeth maintained their hold. As the strength of the frog ebbed and its struggles grew weaker, the snake shifted its grip. Its mouth opened wider and the elastic skin between its jaws stretched. Slowly it began to engulf the frog, working its jaws alternately. The frog's head and forelegs disappeared into the snake's mouth, and its hind legs waved and kicked spasmodically.

The snake rested often as it swallowed the frog, and its head and the forepart of its body bulged and stretched. The bulge moved further back along its body as more of the frog disappeared. Finally only a small part of the hind legs with their webbed feet were visible. Then they, too, slid slowly out of sight. The snake's head returned to its normal size and it opened its mouth several times in gigantic yawns as it re-articulated its jaws.

As the snake moved off into the undergrowth, I could see the frog merely as a large swelling in the region of its stomach. I knew that powerful digestive juices were already at work, breaking it down and supplying nutrients to the snake's body.

Above the spot where the frog had been sitting, a corpulent bumble bee lumbered along in search of flowers. Peace returned to the scene.

General Information

This large group of animals includes about 2,700 species on a worldwide basis. There are about 112 species living in the United States and Canada, 17 in the Northeast. Together with the lizards, they make up the order Squamata.

All Northeastern snakes have elongated bodies, with overlapping rows of scales covering the upper surface and large, overlapping transverse plates on the underside. There is no external ear opening and there are no eyelids. While all species have teeth, only venomous snakes have fangs. There are several modifications for swallowing prey larger in diameter than the snake itself: the two halves of the lower jaw capable of wide separation, elastic skin between the scales of the body, and an elastic gut.

Mating normally takes place in the spring. The female gives off an odor that is left on vegetation as she moves along, and the male follows this scent until he reaches her. Some species have a form of courtship prior to breeding. Some Northeastern species lay eggs; some give birth to live young. Those that lay eggs deposit them in such places as manure piles, sawdust heaps, in or beneath rotting logs, in soft soil, or under rocks. The hatching time is variable and depends to a great extent upon the amount of heat reaching the eggs. When it is ready to emerge from the egg, the young snake uses an "egg tooth" to help cut its way out of the shell. This tooth is shed within a day or two after hatching. Young snakes are small facsimiles of their parents, but in some species they may have a different color pattern until they have grown considerably. In the case of live-bearing snakes, the young emerge from the mother's body surrounded by a thin membrane that they easily break with their heads. There is no placenta as in mammals. Once hatched or born, the young snakes are on their own, for there is no parental care. They are usually able to breed in two to three years.

All snakes are carnivorous. Their food varies greatly, depending upon the species, the size of the individual, the location, and the time of year. Small snakes feed on earthworms, insects, and other small animals. Larger species may eat nestling birds, mice, rats, chipmunks, and the young of squirrels, rabbits, and other small mammals. Those species that live near water may also feed on fishes and frogs.

Most Northeastern snakes are active either by day or at night. They move by throwing their bodies into lateral curves so that the overlapping plates on their undersides push against irregularities on the ground. They are all adept swimmers, and many of them can climb (although most species are seldom seen in trees or shrubs). In the Northeast, only the hognose snake is believed to burrow, but there are some species that may use burrows dug by other animals. While some snakes produce sound by hissing, vibrating the tip of the tail among dead leaves, or shaking rattles (rattlesnakes only), all species are voiceless.

As a snake grows in size it periodically sheds its outer skin. The skin first splits at the snout and then, as the snake moves along, it is gradually peeled off—inside out.

Some snakes (such as the black rat snake) kill their prey by constriction. This is done by wrapping their bodies around the food animal and tightening the coils each time the animal exhales, so that it dies of suffocation. Others (e.g., the black racer) kill by pressing down their bodies onto the victim and thereby suffocating it. Some (including the copperhead) inject venom into the prey through hollow fangs, and yet others (e.g., the garter snake) merely grasp the animal and swallow it alive.

All northeastern species hibernate. The sites chosen may be under rotting logs, beneath rocks, in caves, etc., beyond the frostline.

Snakes are of considerable importance ecologically. The larger species assist in reducing rodent populations, while smaller species eat insects.

There are probably more fables, legends, and misinformation about snakes than about any other group of animals. As they watch a snake's forked tongue flickering in and out of its mouth, some people think it is a "stinger." This is quite untrue. The tongue is used for carrying tiny particles from the air into the snake's mouth. There they are brought into contact with the roof of the mouth, where there is a special structure (Jacobson's organ) that is linked with the sense of smell. This is why a snake is constantly flicking out its tongue; it assists in "smelling." Another common misbelief is that snakes hypnotize their prey. The fixed stare, due to a lack of eyelids, is probably the basis for this story. I have heard it said that mother snakes swallow their young if danger threatens, and later regurgitate them none the worse for the experience. This is most certainly not so. There is no parental care at all, and in any case the very strong digestive juices of the snake would certainly kill the young. Nor is it true that if a snake is mortally wounded it will die only after the sun sets. It dies as quickly as any other animal! How about the common belief that snakes are slimy? This is another falsehood; their bodies are covered with perfectly dry scales.

All of these stories have concerned snakes in general. There are also some tall tales about particular species. The milk snake, for example, is thought to have received its name from an old rural belief to the effect that it sucked milk from cows. Even if this were physically possible—which it is not—no cow would stay still and allow the sharp-pointed teeth of *any* snake to come into contact with her teat! Milk snakes are sometimes found around cow barns, but the reason for their presence is probably that they are searching for mice. The mice are attracted to these buildings by the grains that are fed to cows. It was also once widely believed that the black rat snake (formerly called the pilot black snake) led rattlesnakes to hibernating dens. This belief probably gained recognition when rattlesnakes and black rat snakes were found in the same hibernating sites. As a matter of fact, several species may be found together while hibernating. I once found two ring-necked snakes, five red-bellied snakes, a smooth green snake, and three small northern water snakes all hibernating beneath one large rock. Finally, there is the still prevalent idea that it is possible to tell the age of a rattlesnake by counting the number of segments making up its rattle. It is true that a new button is added to the rattle each time the snake sheds its skin, but it may shed it several times in a year, or not at all. Shedding is largely dependent on how well it feeds. Also, it often happens that, as the rattlesnake moves along, part of the rattle may catch on a thorn or some other sharp object and tear off.

When many people think of snakes they think of venomous snakes. *Some* snakes *are* venomous. That is a fact. It is a fact that often causes *all* snakes to be regarded as venomous. That is ridiculous. Of all the species of snakes in the Northeast, only three are poisonous. A little study of their colors and patterns suffices to make them easily identifiable. The chances of meeting up with any of them are small—there is probably a greater chance of being struck by lightning than by a venomous snake. This is not to say that they should be held in contempt. The wise person always keeps his eyes open, whether in urban or rural surroundings. Most snakes (including nonvenomous species) will bite if provoked, but none will attack a person unless feeling really threatened. They do not chase people (another old belief), and even if this were to occur, numerous tests have shown that even the swiftest American snakes are incapable of moving at much more than five miles per hour, all appearances notwithstanding!

The venomous Northeastern species belong to the pit viper family, so called because members of this group have two small openings (pits) on the snout, in addition to the nostrils. These pits are receptive to heat and assist the snake in locating small, warm-blooded animals. The venom is secreted by glands located inside the head, one on either side of the upper jaw. A duct leads from the gland to the base of each fang. When the snake bites, it opens its mouth very wide and the fangs, which are normally folded along the roof of the mouth, are rotated forward. The venom is injected through the hollow fangs and into the animal. This mechanism is essentially for obtaining food, its defensive function being more or less accidental and merely a side effect of the biting action shown by most snakes when protecting themselves.

Head of a pit viper

In the unlikely event that you or a companion are bitten by a venomous snake, do not try extensive treatment yourself except as an absolute last resort—and even then only if you are positive it was a venomous species! If possible, a tourniquet (a handkerchief or a bootlace will serve this purpose) should be applied at a point between the bite and the heart. It should be loosened every ten minutes or so, to avoid gangrene. With as little movement as possible on his part, the victim should then be taken quickly to a doctor. (Movement causes the blood circulation to speed up.) Cutting the wound to permit the blood to flow and carry out the venom is no longer recommended; the unskilled operator may easily sever a major blood vessel. Suction may be applied to the fang marks, but it must be continued until reaching the doctor. Suction by mouth should be applied only if there are no cuts or sores in the mouth, otherwise the venom will be carried into the blood-stream of the person doing the sucking. In spite of popular belief, do *not* give alcohol to the patient. This also speeds up the circulation!

If bitten by a *non*venomous snake there is little to fear. The two deep puncture marks made by the fangs of a poisonous snake will be missing, but the sharply pointed teeth possessed by all snakes will probably cause a little bleeding. Put a little antiseptic on the shallow wound to reduce the risk of infection. Although some species, if handled, may secrete a foul-smelling liquid from their anal glands, this liquid is not harmful. Merely wash it off. It is a defensive mechanism to discourage other animals from trying to eat them.

Snakes have a bad reputation. They are almost universally regarded as venomous and vindictive, heartless monsters, cold, clammy creatures possessed of almost magical powers—the epitome of evil. Down through the ages, from the Bible to present-day pulp magazines, they have been the subject of all manner of biased writing. Superstition and prejudice continue to follow them. When coming upon a snake—any kind of snake—normally courageous folk shrink in fear.

Yet, when looked at objectively, snakes are merely a specialized group of predators. As such, they are certainly no worse than lions or hawks—or even people, for that matter. Unlike tigers and weasels, they do not kill for killing's sake. They do not play with their victims before killing them, as house cats do. They lack limbs for pursuit or escape, or for holding their food. Therefore, they have evolved other methods for stalking and killing their prey, behavior to scare off intruders, and recurved teeth to hold an animal securely in their mouths. Snakes are neither cunning nor vindictive. When confronted by danger, their reaction is to attempt to get away. If this is not possible they coil and threaten, and if attacked they may bite. Who can blame them? Few animals will suffer themselves to be killed without putting up some sort of fight. But, where snakes are concerned, reason and humanity are forgotten. Ignorance begets terror. Bigotry overcomes understanding.

A snake is a creature of intriguing body form and subtle color. If common sense is used, there is nothing to be feared from it, and much of interest is to be learned. It is merely another, different form of life, another component of the entire country scene.

Keeping Snakes in Captivity

Most snakes are easy to maintain. Depending upon the length of the snake, use an aquarium tank capable of accommodating it comfortably and having a good, well-secured lid. By bracing itself against the floor of the container a snake is able to push up even a fairly heavy lid. Put heavy rocks on the lid as an added precaution.

Since snakes are rather messy when they evacuate their waste, it is best to have no soil or other material that is difficult to clean inside the tank. Simply cover the floor with sheets of newspaper. This is quite absorbent and is easily removed. Even so, the tank will have to be washed out quite frequently. Transfer the snake to another container while doing this.

Put a large, fairly deep dish full of water into the tank, and change the water, if possible, every day. The only other item required is at least one large, rough rock. This will make it easier for the snake to shed its skin by rubbing against the rock.

Food requirements will, of course, depend upon the species of snake and upon its size. Small species will usually take earthworms or small pieces of meat or fish. Larger specimens take larger pieces. Using forceps hold the food near the snake's head and move it slowly from side to side to make the snake think live food is being offered. White laboratory mice are often fed to larger species. If possible, train the snake to take dead mice, so there is no chance of the mouse biting and injuring it. Again, move the dead mouse in such a way as to make the snake think it is alive. Feed any snake until it will take no more, and try to give it food two or three times each week.

Do not keep venomous snakes in captivity; it is too risky unless you are a real expert and have the proper equipment for housing them.

Eastern Worm Snake *(Carphophis a. amoenus)*　　　　PLATE 7

Adult Size. 7″ to 10″; rarely larger

Description. This is a small snake with smooth scales that give it a very shiny appearance. The eyes are very small, and the head is

Eastern worm snake

Northern ringneck snake

Smooth green snake

Eastern hognose snake

young

adult

Northern black racer

Black rat snake

Northern water snake

pointed and no wider than the body. At first glance, this animal might be taken for an earthworm. The upper parts are brown, ranging from a very dark shade to a light, reddish brown. The lower sides and the underside are pink. The tail is sharply pointed.

Breeding. The worm snake lays eggs during July. A captive 10″ specimen laid two eggs, each of which was 1.5″ long. I once found a clutch of four eggs in a damp section of a haystack, but in this case the eggs measured slightly less than 1″. Upon hatching, the young are darker in color than the adults, and measure up to 4″.

Habitat. This is a very secretive snake. It spends much time burrowed under loose soil, beneath stones, under piles of dead leaves or hay, or in recesses within rotting logs.

Food. I have fed captive specimens small earthworms. In the wild it probably also eats insects.

Comments. In my experience, this snake has never attempted to bite. It usually pushes the spine at the tip of its tail against one's hand, but it is not sharp enough to penetrate the skin.

Range in the Northeast. Southwestern Massachusetts, western Connecticut, and southeastern New York (including Long Island) south into New Jersey (Within its range, this species is probably fairly common, but its secretive ways and small size result in its rarely being seen.)

Similar Northeastern Species. The red-bellied snake has three yellowish spots at the rear of the head. The northern brown snake has two rows of dark spots running down the back, with a gray-brown stripe between them.

Northern Ringneck Snake
(Diadophis punctatus edwardsi)   PLATE 7

Adult Size. 10″ to 18″ (although usually about 12″)

Description. This slender, smooth-scaled snake has a yellow ring behind the head. The upper parts are always uniformly colored, but the color may vary from a fairly light gray through bluish gray to quite a reddish brown. The underside is also uniformly colored, but is any shade of yellow from lemon to orange. There is sometimes a row of black dots down the middle of the underside.

Breeding. Three or four eggs (and sometimes up to seven) are laid in early summer, usually in rotting logs or in sawdust piles. The eggs are about 2″ long. They hatch during August or September, at which time the young snake is about 4.5″ long.

Habitat. I have most often found this secretive species beneath flat pieces of shale at the edges of fields that slope down to woods. Other places of concealment have included rotting logs near a beaver pond, an old stone barn foundation, and a rock pile at the edge of a field.

Food. In the wild, this snake includes small frogs and salamanders in its diet, but in captivity I have maintained individuals for long periods by feeding them earthworms.

Comments. Although this snake normally does not bite, the occasional animal will do so when first handled. Because of the ringneck's small size, the bite is insignificant. When first caught it usually ejects a foul-smelling liquid from the anal opening.

Range in the Northeast. Throughout the Northeast, but in "pockets"

Similar Northeastern Species. The red-bellied snake has light spots on the back of the head. The young of the northern brown snake has a light band behind the head, but the scales on the body are keeled.

Eastern Hognose Snake *(Heterodon platyrhinos)* PLATE 7

Adult Size. About 24″, but occasionally up to 3′

Description. The hognose snake is rather heavy bodied, and it has keeled scales. While quite variable in color, it is usually yellowish and brown with at least twenty light bands across the body and a series of large dark blotches down the back. Black, or almost black, specimens are fairly common. The most outstanding feature is the snout, which is distinctly upturned, and is used to burrow in sand or loose soil. The underside is mottled grayish or yellowish, but the tail (the area from the anal opening to the posterior end of the animal) is appreciably lighter in color.

Breeding. Mating occurs in the spring, and an average of two dozen eggs are laid in June or July. (This number varies from seven or eight to more than three dozen.) Newly hatched young are about seven inches long.

Habitat. Since this is a burrowing species, it is usually found where there is loose, dry soil, or in sandy areas. I also know of an abandoned railroad track that used to harbor a fair-sized population of these snakes.

Food. Much of the diet consists of toads, but the hognose snake also eats other amphibians.

Hognose snake

Comments. When disturbed in the wild, this snake puts on a most fantastic defensive display. It first flattens its head and the front part of its body (rather like a cobra spreading its hood) and inflates the rest of the body. The tail is curled tightly. Then it opens its mouth in a gigantic gape, giving it a truly ferocious appearance, and hisses loudly. It strikes repeatedly at the intruder. (This is pure bluff. I have put my hand close to a "striking" hognose and it has closed its mouth and merely banged its head against my hand.) If this does not succeed in scaring away the potential enemy, the snake then writhes around with its mouth open and finally rolls onto its back and pretends to be dead. If picked up, it will merely hang limply without any sign of movement. However, if placed back on the ground right side up it will at once twist over onto its back again, thus giving the game away! Although this is an intriguing display, the hognose very quickly becomes tame in captivity and cannot be induced to repeat its performance. This remarkable display has earned the snake its other name of "puff adder." This is a great pity, for most people are aware that adders (which occur in Eurasia) are venomous, and therefore assume that the puff adder must also be venomous.

Range in the Northeast. Southern New Hampshire, Massachusetts, Rhode Island, Connecticut, and southeastern New York (including Long Island) south into New Jersey and eastern Pennsylvania

Similar Northeastern Species. Although sometimes mistaken for a rattlesnake, the hognose can readily be identified by its upturned snout and its alarming display.

Smooth Green Snake *(Opheodrys vernalis)*   PLATE 7

Adult Size. 13″ to 18″, occasionally to over 20″

Description. This is a smooth-scaled snake, uniform grass green above. The underside is pale yellow or whitish.

Breeding. About seven eggs are laid during late July or August. They measure about one inch and hatch in late August or early September. Newly hatched young measure about four inches in length and are colored dark gray.

Habitat. As one might expect from its coloring, this little snake is usually found associated with grass. I have sometimes inadvertently killed one while mowing my lawn. I have also found them in wet meadows and on roadside embankments. In early spring I have discovered them coiled up beneath large, flat stones on hillsides. One such stone yielded no less than five of these snakes, measuring from eleven to seventeen inches long. Since there were still patches of snow on the ground, these were presumably still occupying their winter quarters.

Food. Insects form the bulk of the diet, grasshoppers being commonly eaten once they become active in the fields.

Comments. This is certainly one of the most docile of all snakes. I have never known one even to attempt to bite.

Range in the Northeast. Throughout the Northeast, although much more common in some localities than in others

Similar Northeastern Species. The rough green snake has been reported in one or two southerly locations in the Northeast, but must be regarded as only an extremely rare possibility. Its scales are keeled and it is more slender than the smooth green snake.

Northern Black Racer *(Coluber c. constrictor)*   PLATE 7

Adult Size. 40″ to 60″

Description. Black racers are rather slender, shiny snakes that are entirely black above and a uniform dark gray on the underside, except for white chins and throats. The scales are smooth.

Breeding. About twelve eggs are laid in manure piles, old mounds of hay, or similar sites during July (in the Northeast). They hatch in late August, the newly hatched young being up to twelve inches long and grayish colored, with a row of brown blotches running down the back but not onto the tail.

Habitat. Although found mostly in wooded areas, this snake is apt to be seen almost anywhere about the countryside. I have found it in sandy areas, in meadowland, and in the vicinity of a beaver pond.

Food. Although almost any small animals are eaten, frogs and mice probably make up a large part of the diet.

Comments. This is a very nervous, excitable species. It invariably attempts to bite, and never seems to become tame in captivity. Although the record length for this snake is given as seventy-three inches, I once had a seventy-eight-inch specimen from Orange County, New York. It was in perfect condition, and I kept it in a cage along with a large black rat snake. In addition to its record size, it was unusual in that it fed well in captivity—it even snatched mice away from the rat snake. As with all members of this species that I have kept, it never became tame, and at one point had a badly injured snout from repeatedly striking at people as they passed its cage. I kept the snake for about a month and then released it into the same area from which it had been captured.

When alarmed, the black racer rapidly vibrates the tip of its tail. If this should occur when the animal is lying among dead leaves the effect is remarkably similar to the sound produced by a rattlesnake!

Range in the Northeast. Southern Maine, New Hampshire, and Vermont; all of Massachusetts, Rhode Island, and Connecticut; New York south of the Adirondack Forest Preserve (including Long Island); and south into New Jersey and Pennsylvania

Similar Northeastern Species. The black rat snake has a thicker, less shiny body, lightly keeled scales, and a whitish underside with dark blotches.

Black Rat Snake *(Elaphe o. obsoleta)* PLATE 7

Adult Size. 48″ to 72″; occasionally to more than 84″

Description. This large, fairly robust snake has plain shiny black or blackish brown upper parts. The scales are slightly keeled, but the keels are difficult to see. On the underside the chin, throat, and foreparts are white; the rest of the belly is whitish with black blotches.

Breeding. About twelve eggs are laid in July, usually in manure piles or soft soil. They measure approximately 1.5″ x 1″. (A captive specimen laid sixteen eggs on 14 June.) The young snakes measure about 12″ when the eggs hatch in the fall. They are colored very differently from the adults: grayish, with a series of large, reddish brown blotches down the back and a row of smaller blotches of the same color down each side.

Habitat. Although I once found a large specimen in a cowbarn, this is mostly a species of wooded areas, and it is found from low elevations up to quite high, rocky hillsides. This snake climbs well, and has been observed fairly high in trees.

Food. In the Northeast, this powerful constrictor is one of the most important reptilian controls on the populations of small rodents. It also takes nestling birds. Apparently, frogs and other cold-blooded animals are never eaten.

Comments. In my experience, the black rat snake can be tamed well in captivity and makes a very good pet. One such pet snake, believed to be about three years old when captured, was maintained for a further sixteen years at The American Museum of Natural History. It was a great favorite with children, and although often handled by them was never known to bite.

Range in the Northeast. Southwestern Vermont, Massachusetts, and Rhode Island; Connecticut; New York south of the Adirondacks Forest Preserve (but not Long Island); and south into New Jersey and Pennsylvania

Similar Northeastern Species. The Northern black racer has a more slender body, smooth scales, and, except for a white chin, a uniform gray underside.

Eastern Milk Snake *(Lampropeltis doliata triangulum)*   PLATE 8

Adult Size. Up to about 30″, but occasionally over 3′

Description. This animal is fairly slender with light gray or buff upper parts marked with large, deep red or brown blotches outlined in black. These are basically in three rows. The largest blotches run down the back, and on each side a row of smaller blotches alternates with those of the back. A light-colored V- or Y-shaped marking is usually present on the head. The scales are smooth, giving the snake a shiny appearance. The underside is white, with a checkerboard pattern of small black squares.

Breeding. The eggs are deposited in warm, moist areas (rotting hay, manure piles, etc.) from mid-June to mid-July. There are usually about twelve eggs. The young have brilliant red blotches, and measure about 8.5″ in length.

Habitat. It is possible to find this species almost anywhere within its range, including wooded areas, meadowland, weed patches, and farm buildings. For several years I have had a number of milk snakes inhabiting the flower bed that runs along the front of our house. There is a concrete walk leading to the front door, and these snakes are able to get underneath it. I believe they may hibernate there.

Food. The favorite food seems to consist of mice. Nestling birds are also taken.

Comments. Milk snakes often bite when first handled. The bite is insignificant, although it may draw blood. If it is available, put a little antiseptic on the wound.

Range in the Northeast. Southern Maine and throughout the rest of the Northeast except northern New Hampshire

Similar Northeastern Species. The northern copperhead has no markings on the head, is colored overall with varying shades of copper, and has crossbands widest on the sides. The northern water snake has heavily keeled scales and reddish, half-moon–shaped markings down the underside.

Northern Water Snake *(Natrix s. sipedon)* PLATE 7

Adult Size. Average about 30″, but sometimes to well over 3′

Description. Heavily keeled scales give this thick-bodied snake a very rough looking skin. The upper parts are a bright reddish brown in younger specimens, but much darker—almost grayish black—in older snakes. A series of whitish or cream-colored bands runs down the body. These are outlined in black, but the black is hard to see in larger snakes. The underside is yellowish, marked with many red (or sometimes grayish) half-moons.

Female water snake and newly born young

Breeding. The young are born alive in late summer or early fall. There are usually about thirty young, but the number varies from about twelve to forty. Newly born young are about nine inches long.

Habitat. These snakes are almost always associated with lakes, ponds, marshes, swamps, and other aquatic environments. A good place to find them is around beaver ponds, where they may often be seen

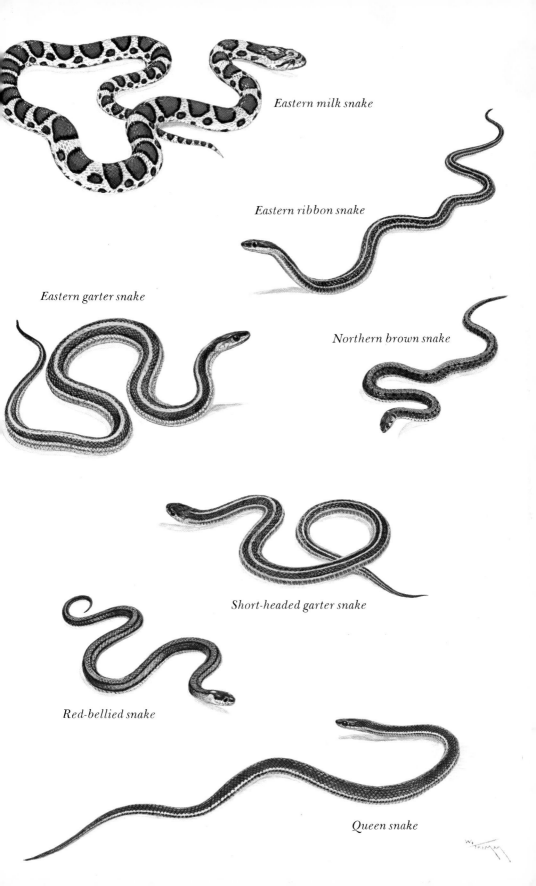

Eastern milk snake

Eastern ribbon snake

Eastern garter snake

Northern brown snake

Short-headed garter snake

Red-bellied snake

Queen snake

sunning while lying fully extended along the branches and logs that have been built into beaver dams and houses. Sometimes a dozen or more may lie together at such a site, with their bodies forming such a tangle that it is difficult to count them. Hibernation sites, however, may be some distance from water.

Food. As might be expected from their habitat, these snakes feed mostly on frogs, fishes, and other aquatic or semiaquatic animal life.

Comments. This species is often confused with the water moccasin (cottonmouth), and thus is greatly feared. It should not be. The moccasin is a venomous species that ranges no farther north than southern Virginia, and that is therefore entirely absent from the Northeast. (Caution! All snakes—including rattlesnakes and copperheads—can swim. Because a snake is seen in the water, do not assume that it is a water snake.)

While not venomous, the northern water snake often bites very viciously when handled, and is large enough to draw blood. Put a little antiseptic on the wound to prevent possible infection. Another defensive action it uses is to discharge a yellow-green, foul-smelling liquid through its anal opening. While messy, this liquid is perfectly harmless.

Range in the Northeast. From southern Maine throughout the rest of the Northeast, except for extreme northern New Hampshire and Vermont

Similar Northeastern Species. The eastern milk snake usually has a light-colored **Y**- or **V**-shaped marking on the head, and a black and white checkered pattern on the belly. The northern copperhead has varying shades of distinctly copper coloring, with the crossbands being wider on the sides of the body than on the back.

Northern Brown Snake *(Storeria d. dekayi)*   PLATE 8

Adult Size. 9″ to 14″

Description. This small, relatively stout-bodied snake has heavily keeled scales. The upper parts are brown, reddish brown or gray, with two rows of small black spots down the back bordering a lighter (usually grayish) stripe. The underside is light brown or pinkish. The top of the head is dark.

Breeding. Mating is in early spring. The young are born alive during late July and early August. One of my captive snakes of this species gave birth to sixteen young, but the usual number is about twelve. The young have a bright yellow collar behind the head and measure about four inches in length.

Habitat. Within its range this little snake may be found almost anywhere, even inside the boundaries of large cities. I once found several beneath stones at an excavation site in New York City, and they are commonly found in vacant lots, gardens, and city parks. Elsewhere, I have usually discovered them hiding beneath logs and rocks in wooded sections, and in stone walls in farm areas.

Food. Earthworms and slugs are probably the most favored food items.

Comments. When handled, these little snakes normally make no attempt to bite. In any case, their teeth are so small that they would probably not break the skin.

Range in the Northeast. Throughout the Northeast.

Similar Northeastern Species. The eastern worm snake has smooth scales and no markings on the back. The red-bellied snake has three yellowish spots at the rear of the head, and a red or reddish belly.

Red-bellied Snake *(Storeria occipitomaculata)*   PLATE 8

Adult Size. About 10″

Description. This little snake is reddish brown, brown, or gray above, with three yellowish spots behind the head. The scales are keeled. There may be four indistinct thin dark stripes running the length of the animal, or sometimes a broad, lighter stripe down the center of the back. As the name suggests, the underside is normally red, but the shade of red varies from very dark to pink. Some individuals may have orange or even yellowish bellies, rather than red.

Breeding. The young are born alive, usually in August, and average from seven to ten in number (although in some instances the number may range from one to twelve). They are only about three inches long at birth, and are darker than the adults. Instead of three well-defined spots behind the head, they have a light collar.

Habitat. This is another species often found lying coiled beneath flat pieces of shale, usually on shallow hillsides, but sometimes just inside wooded areas. Quite commonly I find them beneath the hay mulch on my vegetable garden or lying between bales of hay stacked nearby.

Food. This snake feeds mostly on earthworms and slugs.

Comments. The red-bellied snake is a gentle species, not given to biting.

Range in the Northeast. Throughout the Northeast, but common in many localities and rare or entirely missing from others

Similar Northeastern Species. The northern ringneck has a collar behind the head, rather than three spots. The northern brown snake may have similar-colored upper parts, and the young have a yellowish collar (not three spots).

Eastern Garter Snake *(Thamnophis s. sirtalis)* PLATE 8

Adult Size. Usually about 24″, but occasionally quite a bit larger

Description. This moderately thick-bodied snake has rather heavily keeled scales. The color of the upper parts is extremely variable, ranging from almost black to quite greenish. There are almost always three yellow stripes, one along the top of the back and one down each side. These lateral stripes run along only the second and third rows of scales (counting upward from the large plates that run across the underside.) There are also two rows of dark spots between each lateral stripe and the one along the top of the back. The tongue is bright red. The belly is light greenish or yellow with two rows of black spots.

Breeding. The female gives birth to what may be a surprisingly large number of young during July or August. The average is about twenty, but as many as seventy have been reported, and forty is not at all uncommon. The young are colored like the adults, and measure about seven inches at birth.

Habitat. Almost any type of area may harbor this snake: wet meadows, marshes, woods, garbage dumps, stone walls, gardens, and stream banks, to name but a few. It seems to prefer fairly moist areas.

Food. These snakes feed largely on salamanders, frogs, tadpoles, and large insects.

Comments. Each year I find several garter snakes spending the winter months in a shallow, stone-lined field well on our farm. They may be seen there at any time during the winter, and apparently do not fully hibernate. I have watched them stretched out motionless up to seven inches below the surface of the water in early January. When I have touched them with a stick they moved slowly back between the rocks lining the well.

Northern coal skink

adult

young

Five-lined skink

Northern fence lizard

yellow color phase

Timber rattlesnake

dark color phase

Eastern massasauga

Northern copperhead

Snakes of this species are usually quite pugnacious when picked up, biting fiercely and discharging an odorous liquid through the anal opening.

This snake is often wrongly called the "garden" snake.

Range in the Northeast. Common throughout the Northeast

Similar Northeastern Species. The eastern ribbon snake is much more slender, and its stripes are brighter and more golden-yellow. The stripes along the sides occur on the third and fourth scale rows (counting upward from the belly plates). In the Northeast, the short-headed garter snake is found only in a limited area in southwestern New York. It has no more than seventeen rows of upper scales, whereas the eastern garter snake usually has nineteen.

Eastern Ribbon Snake (*Thamnophis s. sauritus*)   PLATE 8

Adult Size. About 24"; rarely more than 30"

Description. The ribbon snake is slender, with dark brown or black upper parts and three very bright golden-yellow stripes running down the body. The scales are keeled. Two lateral stripes are found on the third and fourth rows of scales (counting upward from the belly plates). The underside is pale green without any markings. The tail is very long, about one-third of the total length of the animal. (On any snake the tail begins at the anal opening.)

Breeding. In the Northeast the young are usually born in August and number from five or six to no more than twenty. They measure about eight inches at birth.

Habitat. This species seems to prefer areas close to water: bogs, ponds, swamps, and marshes. I once found a large, almost three-foot specimen near a rock pile on a hillside some distance from water.

Food. Cold-blooded animals such as frogs and small fishes seem to be preferred.

Comments. Although no snake can move along the ground very rapidly, this species appears to streak along between the rocks and grass when pursued. If chased near water, it will usually plunge in and swim away on the surface. It is a very "peppery" animal, often biting viciously if handled and discharging a bad-smelling liquid through its anal opening.

Range in the Northeast. Not very common in any part of its range, but found in southwest Maine, southern New Hampshire and Vermont, all of Massachusetts, Rhode Island, and Connecticut, and

New York south of the Adirondacks (including Long Island) and south into New Jersey and Pennsylvania

Similar Northeastern Species. The eastern garter snake has less brilliant stripes, and its two side stripes are found on the second and third scale rows rather than on the third and fourth. It also has black spots on its belly.

Northern Copperhead *(Agkistrodon contortrix mokeson)*   PLATE 9

Adult Size. Average about 28″, but sometimes to over 36″

Description. The copperhead is a heavy-bodied snake with keeled scales. The coloring on the upper parts is of several copper or chestnut shades. The head is usually a lighter, almost golden, copper and is entirely without markings. The body is chestnut with a series of darker bands along its length. These bands are shaped more or less like hourglasses, with the narrow part of the hourglass on the back and the widest parts on each side. Some of these markings may be incomplete. The belly is pinkish brown with a dark mottling. There is an opening (the "pit") between each eye and nostril, and the pupil of the eye is narrow and vertical.

Breeding. The female gives birth to about six young in late summer. The young are about 8.5″ long, and are lighter in color than the adults, with the tips of their tails being yellow.

Habitat. Copperheads prefer forested hill slopes with rocky outcroppings. They may also be found in fields close to these areas, and I have seen them coiled in the grass in such places. However, they are much more commonly found in areas where dead leaves form a background into which they blend so perfectly that often they are seen only by chance.

Food. Small rodents and nestling birds are preferred, but this species also eats grasshoppers and other large insects.

Comments. This venomous pit viper is normally very unaggressive, preferring to escape notice by remaining quite still. Herein lies the greatest potential danger to man, for if well camouflaged, or lying just below a rock ledge, it is easy to step onto it or close enough to it so that it may strike. While usually not large enough to inject a fatal amount of venom into a human, this is most certainly a dangerous snake and should be treated with caution. Its bite results in intense, long-lasting pain, discoloration, and swelling.

Copperheads seem to enjoy company, and where one is seen there are usually more nearby. Especially in the spring and fall, they can be found close to hibernating sites in good numbers, and during the winter they will sometimes occupy these dens in company with other species.

Range in the Northeast. Connecticut, Massachusetts, southeastern New York (not Long Island), eastern Pennsylvania, and northern New Jersey

Similar Northeastern Species. The milk snake is often confused with the copperhead, but has a shiny body, markings on the head, and a grayish ground color on the upper parts.

Timber Rattlesnake *(Crotalus h. horridus)*   PLATE 9

Adult Size. 40″ to more than 48″; sometimes over 60″

Description. This is a heavy-bodied snake with strongly keeled scales and a wide, blunt head. It occurs in two color phases. In the more common phase the ground color of the upper parts is mostly yellowish, becoming black toward the tail. The head is yellow. The pupil of the eye is narrow and vertical. A series of somewhat V-shaped dark brown blotches runs down the back and sides. Toward the head these blotches are broken up into three rows. There is an opening (the "pit") between each eye and nostril, and a yellowish rattle (which varies in length) is present at the tip of the tail. In the other color phase the ground color is very dark—almost black—and the only yellow present is a thin band that outlines each blotch.

Breeding. From five to seventeen young are born, usually during late August or early September. They are patterned like the adults but are not as brightly colored. At birth they measure about ten inches in length.

Habitat. Although this snake was apparently widespread at one time, it has been killed off in most of its former areas. Today it is found most often in more remote, rocky, wooded regions, usually at fairly high elevations. I know of several former dens, where many snakes of this species would come to hibernate, but these dens have long since been either abandoned or dynamited out of existence. Where these animals are still present they may locally be fairly common, but such pockets are now few and far between.

Food. The favored food seems to consist of mice, rats, and other small rodents.

Comments. Because of the great amount of venom that a large snake of this species is able to inject, this is potentially the most dangerous of the Northeastern serpents. Nevertheless, there is a great deal of needless fear connected with it. It is an unaggressive species unless cornered, and if reasonable care is taken in climbing over rocky outcroppings, walking through thick underbrush in wooded areas, etc., there is little chance of accidents occurring. (For further information on the rattle, and other aspects of this animal, see the section headed "General Information.")

Range in the Northeast. A few scattered locations in all Northeastern states but Maine

Similar Northeastern Species. The hognose snake has a turned-up snout, a narrower head, and no rattle.

Other Northeastern Snakes

Queen Snake *(Natrix septemvittata)* PLATE 8

The adult length is about twenty-four inches. The scales are keeled, and the body is dark brown with three indistinct dark stripes on the back, a yellow stripe low on each side, and two dark stripes down the center of the underside. The belly is yellow. This snake is found mostly along the banks of streams. Like other water snakes, it feeds on fishes, frogs, and other aquatic and semiaquatic animal life. In the Northeast, this species is restricted to extreme western New York.

Short-headed Garter Snake *(Thamnophis brachystoma)* PLATE 8

The adult snake is about sixteen inches long. Although similar in appearance to the eastern garter snake, it lacks the dark spots usually present between the stripes of that species. The scales are keeled. The head is narrow, no wider than the neck. There are seventeen rows of scales on the body, unlike the eastern garter snake, which usually has nineteen. It occurs in the Northeast only in higher areas near Allegany State Park, in western New York.

Eastern Massasauga *(Sistrurus c. catenatus)* PLATE 9

The adult measures about twenty-four inches long. This is a small, *spotted* rattlesnake. The upper parts are grayish to dark brown, with a series of large, dark spots down the back and two rows of smaller spots down each side. The underside is dark. This species prefers swampy areas, where it may be found coiled on tussocks, but it also occurs in wooded, drier sections. Its Northeastern range is limited to a section of central New York, running approximately from Rochester to Syracuse and south to the northern ends of some of the Finger Lakes.

Lizards

An Escape Story

One afternoon in August I was visiting a friend's farm not far from Bear Mountain, New York. It was one of those summer days when a lazy contentment seems to lie across the land. Sounds were reduced to whispers. Somewhere the sweet, monotonous phrases of a red-eyed vireo's song fell softly through the woods. Grasshoppers were rasping in the meadows. Occasionally a green frog would gulp from a small pond nearby, and off in the distance I could hear the sputter of a tractor. But none of these sounds had any urgency to them; they drifted quietly through the air, pleasant yet scarcely to be noticed.

It had not rained for many days. The leaves hung limply; the grass was dry and crisp. Whenever the wind blew across the fields the grass rustled and small puffs of dust erupted from bare spots. In some fields were rows of brown hay bales, in others shocks of wheat stood in long lines.

A black speck made great, slow sweeps across the sky, and as its circling brought it closer I saw that it was a turkey vulture. Its motionless wings formed a wide, shallow V as the warm air currents held it aloft. As it soared overhead its shadow skimmed across the ground.

In a corner of one of the fields is an old barn. Once it had been used for storing hay, but now it is abandoned. Its roof has partially collapsed, exposing beams and rafters, and the big sliding doors have fallen from their runners and are splintered and broken. Inside the barn, dusty cobwebs decorate the dark corners. Outside, amid a tangle of brambles, lie rusty pieces of farm machinery and a jumble of rotting planks. Behind the barn is a pile of lichen-encrusted rocks.

Near the top of this rock pile I saw a lizard crawl out. Its dark body had light lines running along its length, and its tail was bright blue. Creeping closer, I saw that it was a five-lined skink. It was probably still cool, for it flattened itself against the stone so as to expose as much of its body surface as possible to the warmth. It lay on the warm rock, soaking up the sun's heat, with its eyes almost closed and its tiny claws gripping the stone. Its long tail hung over the edge of the rock. Now and then it opened its eyes suddenly, moved its head, or shifted its position slightly. Then it returned to its immobile, flattened stance. Its eyes became hooded again.

Obviously the skink was unaware of me. It dozed on, and I decided to see if I could catch it. I stepped softly up behind it, taking care not to disturb the weeds. Once arrived at the rocks I stopped for a moment and admired the beauty of the lizard's coloring and the delicate appearance of its fingers and toes against the hard surface of the stone. Then I leaned forward and grabbed.

At the last possible moment the lizard became aware of its danger. It sprang to life and leaped ahead, but as it moved my hand closed over it. It wriggled frantically, and pushed out from between my fingers. I almost lost it. I clamped my fingers together more tightly, but by now all but the tail was free. The little animal gave a convulsive twist and its tail broke off in my hand. With the stump seeping blood, the lizard scampered to safety down among the rocks. I stood, chagrined, and gazed at the tail still writhing in my hand!

Thus, as on several previous occasions, I was treated to an illustration of a wonderful escape mechanism. Many species of lizards have a number of areas in their tails where a break can readily occur. When this takes place the blood vessels at that point are closed off at once and the stump of the tail bleeds for only a moment or two. More wonderful yet is the fact that the cells at the breakage point are stimulated to grow a new tail. Unlike the original, this new tail will have no bones in it, and, in the case of five-lined skinks, will lack the brilliant coloring.

Lizards are by no means common in the Northeast. There are a few areas where pockets of a single species may be found, and in northern New Jersey it is possible to find two of the three Northeastern species in one locality. But lizards certainly represent the smallest group of reptiles in the Northeast. In addition, their numbers seem to be dwindling in some of the places where they were once plentiful. To some extent this is due to changes in their habitats. Land once suitable for them has now gone under the plow or has been paved over. Sometimes their scarcity is the result of overcollecting. Today, the sighting of a lizard in most Northeastern areas is becoming more and more of an event.

General Information

Although there are about 3,000 species of lizards living today, with 90 of these found in the United States and Canada, there are only 3 species living in the Northeast. With their relatives the snakes, they make up the order Squamata.

Our Northeastern lizards have elongated bodies, fairly long tails and four legs. They have an outer covering of scales, external ear openings, eyelids, and claws on their fingers and toes. While they have teeth, the teeth are very small. The males are usually larger than the females.

All three of our species hibernate, often by burrowing down into the soil but also in rotting logs and under rock piles and other protected sites. Once they emerge from hibernation, in the spring, the males establish territories and defend these areas against other intruding males. Mating usually takes place two or three weeks from the time they leave their winter quarters. The eggs may be laid at any time from May through August, but most often in late May or early June. They are deposited in places such as soft soil or sand, moist sawdust, manure piles, or decaying logs. Female coal skinks and female five-lined skinks usually protect their eggs until they hatch. Depending upon the location and the species, hatching takes place in four to ten weeks. The baby lizards grow quite rapidly and are able to breed at the age of three years.

The Northeastern lizards are all carnivorous, feeding upon insects, spiders, and other small invertebrate animal life.

Like their close relatives the snakes, they are voiceless, but unlike snakes they are not active at night. Although not often seen in the water they are good swimmers, and when swimming keep their arms at their sides.

The skin is shed periodically, but not in one piece. Often the sections of shed skin are eaten by the erstwhile owner.

If a lizard is attacked, one defensive measure is to shed the tail, which continues to move and may draw the attention of the predator away from the lizard itself. It is thought that the brilliant blue tail of a young Northeastern skink may attract a pouncing predator to the hind end. If the tail is grabbed, it is at once shed and its owner often gets away. Sometimes, when a tail is shed incompletely, or when an animal bites the side of a lizard's tail, the cells in the injured area are stimulated to grow a new tail. The new tail grows from the side of the original one, resulting in a "fork-tailed lizard." Such creatures are often thought to be poisonous. This, of course, is not the case. The only poisonous lizards in the world are the Gila monster, a twelve-inch-long animal found in southwestern United States, and a close relative of the Gila monster named the Mexican beaded lizard.

While Northeastern lizards will often bite when handled, there is nothing to fear from them. They have teeth that are so small that usually they do not even break the skin.

These are fast-moving animals; they can streak along for short distances. Even so, they are often taken unaware, and are eaten by skunks, raccoons, and even some hawks.

Keeping Lizards in Captivity

Use a five-gallon aquarium tank as a container. The environment must be dry, so cover the floor of your container with dry sand to a depth of about three inches. Half-embed some rocks in the sand, and put in some pieces of bark or wood as hiding places for the lizards. To make the habitat attractive you can buy some small potted cactus or other plants and sink the pots into the sand. Put each pot into a glass container so that, when watering, surplus water will not run out into the sand.

To provide heat, suspend a 100-watt light bulb over and near the tank. While it is a good idea to keep this bulb lighted throughout the day, room temperature seems to be adequate for the lizards during the night.

In giving water to your lizards, the best method seems to be to flick water onto the inside of the container walls, or onto the leaves of the plants. As the droplets run down the glass or sparkle on the plants, the lizards will dash over and drink. They usually will not drink from a dish. Be careful not to drench the area!

Feed the lizards with small pieces of earthworm or raw

meat, or with mealworms. (Do not feed too many of the latter; the lizards may develop obstructions in their intestines due to the tough skins of the mealworms.) When offering food, try to convince the lizard that the food is alive by moving it gently back and forth near the animal's mouth.

Five-lined Skink *(Eumeces fasciatus)*   PLATE 9

Adult Size. Total length 4.5″ to 6″; occasionally up to 7″

Description. The tail of this slender lizard is approximately twice as long as the head and body combined. The scales are tight to the body and difficult to see. There are obvious ear openings. In the immature animal the upper parts are very dark, with five light lines running from the snout down the length of the body. The tail is a brilliant blue. As males of this species increase in age they become progressively browner or grayer and lose the blue coloring of the tail. The light lines down their bodies become indistinct, but there is a broad, dark brown or black stripe on each side. The head, especially near the snout and throat, may develop rust-red coloring. Adult females also lose the blue from their tails but have no red on the head. These animals can run very rapidly.

Breeding. Mating takes place from late April to early June. The eggs are laid from June to mid-July and usually hatch, in the Northeast, during August. They measure about 0.5″ x 0.25″, but the size increases as they absorb moisture, and they may eventually measure 0.75″ x 0.5″. They number up to about eighteen, but smaller females may lay only three or four eggs. The site selected is often in or under a rotting log. The female remains with the eggs until they hatch.

Habitat. This species is usually found in open areas surrounded by trees, and where there are rocks or tree stumps to offer shelter and hiding places. I know of one location where these lizards live in a small picnic ground at the edge of some woods, another location where there is a population of them on the exposed top of a rocky hill, and yet another location where they are living in pine woods with loose, very sandy soil. Sometimes they may also be found in fairly damp areas.

Comments. Five-lined skinks live mostly on the ground, but they may bask on logs or other elevated sites. When captured they will usually bite, but normally their teeth are too small to break the skin.

Range in the Northeast. Scattered distribution throughout most of Massachusetts, Rhode Island, Connecticut, eastern New York (but not in the north), and south into New Jersey

Similar Northeastern Species. The coal skink has four light lines running down the length of the body and extending out onto the tail, and a broad, dark stripe along each side.

Northern Coal Skink *(Eumeces a. anthracinus)* PLATE 9

Adult Size. Total length 4.5″ to 6″; occasionally larger

Description. This is a smooth-scaled, shiny, slender lizard with a tail about twice the length of its head and body. (The tail is shorter in young animals.) Its color is basically yellowish brown, with four light lines running from just behind the eyes down the body and out along the tail. These are arranged so that there is one on each side of the back and one halfway down each side of the body. Between the upper two lines (down the top of the back) the coloring is grayish brown, while along the sides the lines enclose a very broad, dark brown or black stripe. The newly hatched young have blue tails and are either striped like the adults or almost entirely black.

Breeding. These animals breed in late May or early June. The only eggs I have ever found were a clutch of seven in a depression beneath a pile of rocks, but the number is usually reported as being eight or nine. The ones I found averaged slightly less than 0.5″ long by 0.25″ wide. Like most lizard eggs, they would absorb moisture and increase in size half again as much before hatching. The female remains with the eggs and guards them. They hatch in about a month.

Habitat. Coal skinks usually live on hillsides where there are trees and rocks.

Range in the Northeast. Found only in parts of western New York, mostly in the Finger Lakes region

Similar Northeastern Species. The five-lined skink, as its name suggests, has five lines running down the body, rather than four, as on the coal skink.

Adult Size. Total length 4.5″ to 6″; occasionally larger

Description. This is the only lizard in the Northeast with rough, spiny-looking scales. The tail is about 1.25 times as long as the head and body. The male has brownish upper parts—sometimes quite a dark brown—with six to ten thin, wavy, dark bars across the body. These bars are only dimly visible, and sometimes may be absent altogether. On his underside there is a large, bright, iridescent dark blue or greenish patch in the throat region, and a similar patch on each side of the belly running from the armpit to the groin. These patches are bordered with black. The rest of the underside is whitish. The female is usually much grayer on the upper parts, and the crossbands are much more distinct. Her underside is white with a sprinkling of small black dots.

Breeding. Mating takes place in April and May, and the gestation period is about eight weeks. Eight eggs are usually laid, but up to seventeen have been reported. The eggs are white and measure about 0.5″ by slightly more than 0.25″. The young lizards are about 1″ long upon hatching. They are sexually mature in two years.

Habitat. Fence lizards prefer dry, often sandy, wooded areas. In such sites they may be found along hedgerows or on timbered slopes.

Comments. As with many other species of lizard, the males establish territories and defend them against males of their own species. (They may also attempt to scare off other types of intruders. I once discovered a fence lizard who considered his territory to be a pile of old planks. Upon seeing me approach he rushed out along one of the planks and proceeded to "flash" his blue throat patch at me. He accomplished this by standing with his forelegs stiff and straight, and then alternately lifting and dropping his head so that the throat patch repeatedly showed and then disappeared. As I moved around the planks he shifted his position so that he was always facing me, and repeated his threat display.)

Male fence lizard "flashing"

Fence lizards often climb trees in order to escape potential enemies, keeping the trunk between themselves and the danger. Like most lizards, they can run very rapidly.

Range in the Northeast. Present only in one or two areas in southeastern New York (including Staten Island, but not Long Island), and in northeastern Pennsylvania

Similar Northeastern Species. None. This is the only rough-scaled lizard in the Northeast.

Mammals

Today's Ruling Class

The edge of the woods at Howbourne

ON A WORLDWIDE BASIS there are about 3,200 species of mammals. Of this total, about 320 land mammals live in the United States and Canada, with 63 occurring in the Northeast.

All mammals have hair over part or most of their bodies at some period during their lives, and all female mammals possess mammary glands that produce milk for feeding the young. Mammals are warm-blooded (their body temperature varies only slightly) and breathe by means of lungs. There is a muscular diaphragm between the chest and abdominal cavities, and they have seven neck bones. (The only exceptions to the latter are sloths and manatees, which are not found in the Northeast.)

Some mammals have limbs and other parts of the body modified for burrowing (moles), swimming (otters), running (foxes), gliding (flying squirrels), flying (bats), climbing (squirrels), or leaping (jumping mice). Tails, if present, may be used for balance (squirrels), swimming (muskrats), gripping (opossums), fly swatting (deer), or steering (beavers). Even the hair serves a variety of purposes: for warmth and camouflage in most species, for protection (porcupine quills are modified hairs), for protecting the soft underfur (as in the outer "guard hairs" of beavers), and even as warning signals to companions (as in the white underside of a deer's tail, which "flashes" when the deer lifts its tail as it runs). The types and number of teeth vary according to the food eaten.

Except for a few egg-layers in the Australian region (duckbilled platypus and echidnas), all mammals are born alive. In most species mating is restricted to definite breeding periods, but in some it may occur at any time. The gestation period (length of pregnancy) is extremely variable, depending upon the species. The young may be born blind, naked, and helpless (altricial) or fully furred, open-eyed, and capable of moving around almost at once (precocial). Parental care in mammals is highly developed. In some species the lifespan is no more than a single year; in others it may be upward of fifty years.

Depending upon their adaptations to various modes of life, different species of mammals may feed on insects and other invertebrates, fishes, amphibians, reptiles, birds, other mammals, dead animals, or plant material. Some mammals are entirely meat-eaters (carnivorous) or insect-eaters (insectivorous), some are strictly plant-eaters (herbivorous), and others feed on a combination of both plants and animals (omnivorous).

Unlike birds, which are mostly active during daylight hours, often colorful, and usually with voices that draw our attention to them, most mammals tend to be largely nocturnal, secretive, not brightly colored, and mostly silent. Thus, although there are mammals in virtually all types of habitat throughout the Northeast, they are often overlooked.

Although they are not usually noisy, most mammals have a well-developed voice box and are able to make sounds. These sounds run the gamut of squeaks, whistles, chirrups, snarls, grunts, growls, and barks, depending upon the species. Such sounds are used for communicating with each other, warning or scaring, finding mates, and, in the case of bats, echo location for avoiding obstacles in the dark.

Some mammals establish territories in which they raise their families and search for food. They usually defend these areas against other members of their own species. The territory may cover many square miles or it may be limited to less than an acre—again, it depends upon the species and upon what it eats.

Although not as common—or as evident—as in birds, migration also occurs in some mammals. In the Northeast this annual movement is restricted to a few species of bats. (See the information under individual species.)

All amphibians and reptiles in the Northeast hibernate. This is not the case with Northeastern mammals, only three groups of which are true hibernators (the jumping mice, some species of bats, and the woodchuck).

Wild mammals have always been important to man. Sometimes this importance is beneficial to us, sometimes it is detrimental; it seems to depend, to a great extent, on what a particular animal is doing at a particular time, and where it is doing it. Meadow mice, for example, eat many weeds and their seeds, and this is certainly of help to a farmer. On the other hand, they also do much damage to farm crops and to vegetable gardens. Foxes eat hundreds of meadow mice, and we would say that is to their credit, but they may also occasionally steal chickens. That is a different matter entirely! In building dams across streams, beavers may prevent flooding from taking place further downstream, but if they build their dams in the wrong place they may actually *cause* flooding of highways or valuable stands of trees. The tunneling and burrowing activities of some mammals help to aerate the soil and make it more fertile, but they may also damage gardens or golf courses. So, in many cases, the value or harmfulness of a mammal is a very relative matter.

Even with the advent of synthetic fibers and the continued use of wool, many mammals are still trapped for the fur trade. Some of these fur-bearers are now rare, and the trapping of them is now strictly controlled, but muskrats, raccoons, beavers, foxes, and others are still trapped in great numbers each year.

Whether or not we may agree with it, hunting for sport is very popular, and huge amounts of money are spent annually by hunters for transportation, special clothing, arms and ammunition, and licenses. Some of the hunted animals are used for food.

Some mammals, or the parasites that live on or in them, are capable of transmitting serious diseases to man. These diseases include rabies, typhus, tularemia, anthrax, and, in the case of the Norway rat, bubonic plague.

There are important agricultural pests among the mammals. Many rodents, including rats, mice, squirrels, and woodchucks, and other mammals such as rabbits, deer, and raccoons, feed on farm crops, either in the field or after they have been harvested. Deer, rabbits, and porcupines sometimes cause problems in valuable orchard or forest areas by "girdling" trees. (That is, they nibble off the bark around the entire trunk of the tree. This destroys the tree's ability to obtain nourishment, and it dies.)

Foxes, bobcats, weasels, and other meat-eaters were once branded as "bad" animals because they hunted down other creatures. It is now realized that all predators are valuable natural controls on the populations of other animals. Without these predators we might well be overrun by animals that we consider to be pests.

Although there are many hair-raising accounts of mammals attacking people, such accounts are almost always either greatly exaggerated or are figments of the imagination. Wild mammals are normally scared of people. They avoid them whenever possible. It is true, however, that under certain circumstances some mammals might be dangerous. If an animal is sick—and particularly if its brain is affected, as with rabies—it will not behave normally and might attack. If a female mammal such as a bear sow feels her young are in danger, she might well lose her fear of man and charge. Wounded mammals will sometimes turn on people. In some species, including members of the deer family, the males become very pugnacious during their breeding seasons and may assume that almost any moving object is a potential rival. Some of the larger wild mammals, particularly bears that have been fed by man and that have, therefore, lost their fear of him, may become dangerous if they scent food at a campsite. However, these instances where attacks might be made are all examples of atypical behavior on the part of the mammal. They occur but rarely. One is much more likely to be attacked by a domestic dog than by a wild mammal.

Keeping Mammals in Captivity

With very few exceptions, wild mammals are protected by law. Some may be subject to a limited hunting or trapping season, but it is illegal to keep even those species in captivity.

Marsupials

Death by Moonlight

The long summer day is over, but night has brought little relief from the heat. There is no breeze, and under a misty full moon the leaves hang limply from the trees. Among the trees, near the shimmering waters of Queechy Lake, near Canaan, there is a clearing dotted with the dark shapes of many picnic tables. A few hours earlier this had been a scene of noisy, festive activity, but now the people are gone, the tables abandoned. Fireflies twinkle at the borders of the clearing. Near the trees are some litter baskets, now filled with paper wrappings, half-eaten sandwiches, apple cores, and other remnants of the day's feasting.

For some time I have been sitting quietly amid a bed of ferns just within the edge of the woods. The litter baskets are only a few yards from me and are flooded with moonlight. Periodically the sonorous notes of a bullfrog float to me from somewhere across the lake—a strangely restful sound.

From the woods I hear a faint rustling. A minute or so later, quite close to me, the fern fronds quiver and an animal comes into view. At first I think it is a large rat, but as it ambles slowly from the shadows and out into the moonlit clearing I see a long white face and realize that it is an opossum. It is not a large opossum, not yet fully grown. Evidently it has visited this picnic area before, for without pausing it makes its way to one of the litter baskets. Here it stops briefly and sniffs through the mesh sides. It is apparently happy with the odors it has discovered, for now, with its long naked tail shining whitely in the moonlight, it clambers up the side of the basket and onto the overflowing contents. Here it begins to search for scraps of food. As it moves around, the rustle of disturbed paper and the occasional clink of empty bottles come clearly to me through the still air.

I sit and think about this creature—a marsupial whose ancestors were contemporary with the last of the dinosaurs and whose living relatives include kangaroos and other pouched mammals. Most of the surviving members of its group live in Australia, which scientists believe became isolated from other land masses before any mammals *but* marsupials had settled there. Elsewhere they fared badly in competing with other kinds of mammals, and, thus, except in Australia, there are few species left. Yet, of these few, our North American opossum seems able to hold its own well enough, and even appears to be steadily extending its range. Perhaps it is more adaptable. Certainly it has become used to living near man. The small animal now rooting about in the litter basket is assuredly taking advantage of what man has to offer. It is literally eating the crumbs from his table!

The swooping of a shadow from the trees interrupts my thoughts. A dark figure slams onto the litter basket. There is a violent struggle, a beating of great wings. A can clatters as it drops to the ground. Paper flies in all directions. Then silence. Looking huge in the moonglow, a great horned owl crouches with its wings covering the opossum. It looks up from its motionless victim and turns its head slowly to survey the clearing. Suddenly it leaps into the air, and with labored wing-beats lifts out over the treetops. As its silhouette appears against the sky I can see the limp form hanging below it.

General Information

The opossum is only one member of the order Marsupialia, which also includes kangaroos, bandicoots, and wallabies. Females of this order normally have a pouch. The young are born after a very short gestation period, and at birth are still a long way from being fully developed. When ready to give birth, the female licks her fur from the vaginal opening to the pouch. After they are born, the young climb along this wet fur and into the pouch. They remain there until their development is completed.

Adult Size. Total length about 31″ (head and body about 18″; tail about 13″); weight about 10 lbs.

Description. This is a rather chunky looking mammal with short legs, a long pointed head, large naked ears, and a long, thin, naked tail. The body fur is coarse and grayish in color, with long white hairs overlying the darker fur beneath. The face is white, the ears mostly black, and the tail is black near the body and whitish throughout the rest of its length. The first toe of the hind foot has no claw and is opposable (it can be used for grasping). This is the only North American land mammal with fifty teeth.

The opossum is mostly nocturnal, but occasionally can be seen in the early evening. It can climb well, and its tail is prehensile (capable of being used for gripping) to assist it. When attacked, it will often pretend to be dead—thus the expression "playing possum." It does not hibernate, but may spend a week or more in its den during very cold weather. If it does emerge during subfreezing weather its ears and tail are subject to frostbite.

Breeding. In the Northeast the young are usually born early in the spring. The gestation period is only twelve or thirteen days, and the tiny, pink young are born naked and with only partly developed eyes, ears, and toes. Up to fourteen or fifteen may be born, although the female usually has only thirteen nipples. As each young one climbs into the pouch it attaches itself to a nipple. Surplus young die. The young remain in the pouch for about two months. Presumably the weaker ones also die, for there are usually no more than eight young by the time they are ready to leave the pouch. As they get older they ride on the female's back by gripping her coarse fur. They leave her when they are about three months old.

Habitat. Although preferring woods, swamps, and farmland, the opossum is apt to turn up almost anywhere. Some friends of mine who live in the heart of a large town on Long Island once found one in their garage, and there are numerous sightings in other built-up areas.

Food. This is an omnivorous animal, feeding on almost anything that is available. Its food includes mice, frogs, earthworms, small snakes, fruits, and berries. It is also a constant visitor to garbage cans and dumps.

Economic Importance. The opossum is still trapped for its fur, which is used mostly as trimming. Few pelts are of good enough quality to be made into coats, but better quality pelts are sometimes dyed, and the results can be quite attractive. Although sometimes used as food, opossum meat is very fatty and oily. This mammal is undoubtedly of some value in rodent control, but it can do considerable damage in a corn or berry patch.

Comments. Where it is common this is certainly one of the animals that suffers most from road mortality. On one fifty-mile stretch of the Taconic Parkway, in New York, I once counted nine opossums that had been run over.

Range in the Northeast. Seems to be constantly extending its range northward; now found throughout the Northeast except for Maine and northern New Hampshire and Vermont

Similar Northeastern Species. While young opossums may bear a slight resemblance to rats, there is really no other Northeastern mammal that looks like them. Simply look for the coarse, grayish fur, a long white face, and a long, naked tail.

Shrews and Moles

A Monster in Miniature

At one corner of our farm is a dense stand of white pines. The lower branches are brown and dead, barren of needles. They died many years ago, for little light reached them through the thick green canopy above. As the trees grew taller, and raised this canopy ever higher, more branches were denied sufficient light, and so they died in their turn. Today these dead, interlacing limbs and twigs are almost impenetrable. One must literally crawl on hands and knees between them.

Because of the thick mat of fallen pine needles there are few plants on the floor of these woods. Here and there mosses and lichens adorn the rotting, papery stumps of toppled trees, and in late summer mushrooms appear. But there is little else, and since the plant life is so sparse there are few animals. Yet, in spite of its somewhat somber aspect, I am fond of this area. On a sunny morning in late spring it is pleasant to creep in beneath the dead, brittle branches and stretch out upon the spongy carpet of pine needles. There is a warm fragrance that is sometimes so strong one can almost see it, a delicious, clean aroma that lingers for a while on one's clothing and, later, in one's mind.

Running through these pine woods is a low stone wall, mute evidence that at one time this was open farmland. Where the sunlight strikes the wall I have sometimes seen garter snakes basking, and have found their shed skins there.

While sitting near this wall one morning I noticed a small mammal scurrying among the fallen rocks. At first I took it to be a deer mouse, for they are common here, but when I moved closer it proved to be a short-tailed shrew, busily searching for food. As it skittered along it would frequently lift its little snout skyward and swivel it in a circle. At such times it managed to look remarkably like a tiny grizzly bear testing the wind for either food smells or potential danger.

Shrews are extremely high-strung creatures, agitated-looking indeed. Their pace is a constant trot. They are always on the move; constantly eating in order to replenish the strength they waste in their frenzied activity, they rarely sleep, and then only in brief snatches. They are active all year long. This frenetic pace may reduce their lifespan to little more than a year. Many do not last even this long, for they are killed by a wide variety of predators. I suspect they are usually killed in error (perhaps being mistaken for mice), for they have powerful skin glands that apparently give them a bad taste. Farm cats often catch them, but they seldom eat them. It must take a strong stomach indeed to make a meal of one.

Shrews are also very pugnacious. They will eat almost any living thing they can overcome. The short-tailed shrew can even kill and consume mice larger than itself—and does not turn up its pointed nose at members of its own species. Its saliva contains a toxin that causes paralysis and even death to animals as large as a mouse, and that has been known to cause serious discomfort to humans.

As I watched my short-tailed shrew it pounced upon a large beetle, and with startling violence proceeded to tear it to pieces. Faint crunching noises reached me. In less than a minute it had bolted down the unfortunate insect. Then, its snack finished, the shrew at once resumed its rushing about in search of more food, and finally disappeared between two rocks in the wall.

I walked home. As I walked I thought about the demonstration of ferocity I had just seen. Wild animals normally give people a wide berth, but I wondered how a shrew might react if it were larger. If this venomous, choleric creature were four feet long, instead of a mere four inches, it would be a formidable animal indeed. In fact, it would be downright nightmarish!

General Information

Shrews and moles, which are included in the order Insectivora, are all small in size, some of the shrews being among the world's smallest mammals. Most have long, pointed snouts. The teeth are pointed and sharp and, as is suggested by the name of the group, are basically adapted for a diet of insects. There are five digits on each foot, and in most species the eyes are tiny. Most are nocturnal.

Masked Shrew *(Sorex cinereus)*   PLATE 11

Adult Size. Total length about 3.5″ (head and body about 2″, tail about 1.5″); weight less than a dime!

Description. The body is covered with very dark brown, almost black, or gray-brown fur. The fur is darker in summer than in winter. The underparts are lighter in color, usually a fairly light gray. The tail is brown above and buff below. The eyes are very small and the ears are hidden in the fur of the head. The tips of the teeth are reddish brown.

This shrew is extremely active both day and night. Its high metabolism—its heart beats more than 1200 times per minute—means that it must constantly search for food, and it sleeps only in short snatches. A very high-strung, nervous, and pugnacious creature!

Breeding. The female masked shrew produces two or three litters per year, from spring until fall, with about six young per litter. She builds a ball-shaped nest of leaves and grass, often under a log or in a shallow hole. The young acquire their fur in about two weeks, and their eyes open in about three weeks. It is thought that these shrews begin to breed at the age of five months. Probably because of their frenetic activity, they have a very short lifespan, usually no more than a year.

Habitat. This shrew can be found in almost any type of area, from marshes and deciduous forests to fields and evergreen woods. I have several times found them in mousetraps in my house near the Berkshires.

Food. Insects make up the main part of the diet, but other small animal life such as earthworms and snails are eaten by the masked shrew. All shrews have enormous appetites, and are said to eat more than their own weight each day. Quite frequently I have found that mice caught by my traps had large holes in their skulls. This was a puzzle until I found trapped masked shrews in the same area. From the fact that I bait the traps with peanut butter and rolled oats, I assume that these little shrews will eat those items also—not to mention the mice.

Comments. Since most of their diet consists of insects, shrews undoubtedly are of some value in the control of harmful insects. Although caught and killed by a host of predators that perhaps mistake them for mice, shrews secrete a strong musky odor that prevents many of these predators from eating them.

Folklore. At one time in Europe it was thought that if a shrew ran across a farm animal that was lying down the animal would suffer intense pain. To counteract this, a shrew would be walled up in an ash tree (a "shrew ash") and then a twig taken from the tree would be brushed onto the suffering animal to relieve the pain. Also, a loud, quarrelsome woman used to be called a shrew, probably because of the shrew's pugnacious nature. The ancient Egyptians thought the shrew was the spirit of darkness.

Range in the Northeast. Throughout the Northeast

Similar Northeastern Species. There are several small shrews that resemble the masked shrew, and it usually takes a real expert to identify any one of them. The smoky shrew is a little larger and is darker on the underside. The pygmy shrew is slightly smaller and has a tail only vaguely bicolored. The least shrew has a much shorter tail.

Short-tailed Shrew *(Blarina brevicauda)*   PLATE 11

Adult Size. Total length about 4.5" (head and body about 3.5", tail about 1")

Description. The body of this large shrew is very thickset. The legs and tail are short, there are no visible ears, and the eyes are minute. The fur is blue-black in color, lighter on the underside. The tail and feet are dark gray, and the teeth are tipped with reddish brown. Like other shrews, the short-tailed shrew is active by day or night at all seasons of the year. Unlike most other shrews, it is able to burrow through soft soil or snow, although it usually does this close to the surface. Because of its larger size, this species is able to inject enough of its poisonous saliva (which most shrews possess) to slow down, or even kill a mouse. If a person is bitten by this shrew, it can be a very painful experience.

Breeding. Two or three litters per year are produced, usually in spring or late summer and early fall. Other breeding habits are similar to those of the masked shrew.

Habitat. This species may be found almost anywhere: marshes, hedgerows, woods, farmyards, fields, and in and around rural homes.

Eastern mole

Hairy-tailed mole

Star-nosed mole

Opossum

Woodchuck

Mountain lion

It is very common around my house, particularly in winter, when I frequently see individuals emerging from their runways under the snow.

Food. In addition to eating insects, this shrew feeds on small mice and salamanders. I often find nothing but a few bones and shreds of fur where a short-tailed shrew has eaten almost an entire mouse from one of my traps. It also eats seeds, and I have watched as many as four of these shrews feeding on sunflower seeds that have fallen from my bird feeders. They must also be attracted by the peanut butter and rolled oats with which I bait my mouse traps, for I sometimes find one that has fallen victim to a trap.

Economic Importance. See masked shrew.

Range in the Northeast. Throughout the Northeast

Similar Northeastern Species. The only other Northeastern shrew with a short tail is the least shrew, which is only about three inches in total length, brown in color, and not present in the New England states.

Star-nosed Mole *(Condylura cristata)*   PLATE 10

Adult Size. Total length about 7.5″ (head and body about 5″, tail about 2.5″); weight about 1.75 ozs.

Description. This animal gets its name from the ring of twenty-two pink, fleshy, tentaclelike structures surrounding its snout. No other mammal in the world has such a peculiar-looking snout! The upper parts of the body and tail are very dark brown or black, the underparts lighter. Its tail is long, hairy, and, especially in winter, rather swollen-looking. Its front feet are very broad (for digging), but not as much so as in other moles. Its eyes are tiny.

Unlike other moles, the star-nosed mole is quite happy in water and obtains much of its food by entering ponds and quiet streams to search for aquatic insects. It swims well, using its large front feet as well as its tail. Like their relatives the shrews, moles are active by day or night; but whereas the eastern mole rarely ventures above ground, the star-nosed mole will often be found at the surface. Its tunnels may also appear beneath a log, with the underside of the log acting as the tunnel ceiling.

Breeding. From three to six young are produced in a single litter each spring. The gestation period is about six weeks. The young are born without fur, but are well covered three weeks later. Soon afterward they leave the rounded, leafy, underground nest to fend for themselves, and are able to breed at the age of about ten months.

Habitat. In spite of its burrowing habits this mole is found in moist areas: wet meadowland, damp woods, and similar areas. I have twice found individuals in a muddy drainage ditch when the water was low.

Food. This animal feeds mostly on insects (including aquatic species), earthworms, and other invertebrates. It may locate food by means of its snout tentacles. Like all moles, it consumes many insect larvae as it moves through its tunnels and runways.

Economic Importance. In eating harmful insects and their larvae this animal is obviously beneficial, although it almost certainly eats just as many insects that we would consider to be helpful. Its digging activities aerate the soil, but sometimes the excavated soil is thrown up on lawns and golf courses.

Range in the Northeast. Throughout the Northeast in suitable locations

Similar Northeastern Species. Both the eastern mole and the hairy-tailed mole have broad front feet for digging, and bodies similar in shape to the star-nosed mole. But they both have short tails, and neither of them has projections surrounding the snout.

Eastern Mole *(Scalopus aquaticus)*   PLATE 10

Adult Size. Total length about 6.5″ (head and body about 5.25″, tail about 1.25″); weight about 1.5 ozs.

Description. The upper parts have tight, velvety fur that is blue-gray in winter and lighter—even brownish—in summer. The underparts are grayer. The tail is short and naked. The front feet are very broad (wider than long) with the palms turned outward and with strong claws. There are no visible ears, and the eyes are minute. (It is thought that perhaps this animal can distinguish only between light and dark.) Moles are active all year long, both day and night.

Breeding. There is a single litter each year, born in April or May. The two to five young that are produced are at first naked, but have grown a good coat of fur by the time they are about a month old. The nest is built below the surface of the ground at depths of up to twelve inches, and is constructed of leaves and grasses. Young moles are mature when a year old.

Habitat. This mole lives most of its life underground, preferring loose, well-drained soil. It is found mostly in open areas such as fields and meadowland, but also occurs in thinly wooded areas. The main tunnels are well below the frostline at a depth of more than twelve inches, but during the warmer parts of the year other runways are made from here to just below the surface of the ground. The excavated soil appears as "molehills," which are usually three or four inches in height. This mole may dig a new tunnel at the rate of fifteen feet per hour. (This is roughly equivalent to a six-foot man taking only an hour to burrow his way for half the length of a football field!)

Molehill

Food. While earthworms and insects and their larvae form the bulk of the diet, some plant material may also be eaten.

Economic Importance. This species' main value to man is in eating harmful insects and their larvae. It is one of the most important natural controls on the larvae of pests such as rose-chafers, Japanese beetles, and Asiatic garden beetles. Its tunnels help to make soil more fertile by aerating it, but unfortunately its mounds of excavated soil sometimes disfigure lawns and golf courses. When the latter happens, steps are often taken to trap or poison these moles.

Range in the Northeast. Southern Massachusetts and Rhode Island, Connecticut, southeastern New York south of the Catskills (including Long Island), eastern Pennsylvania, and northern New Jersey

Similar Northeastern Species. The hairy-tailed mole has a short but hairy tail, and the star-nosed mole has a long tail and a ring of twenty-two tentaclelike structures surrounding its snout.

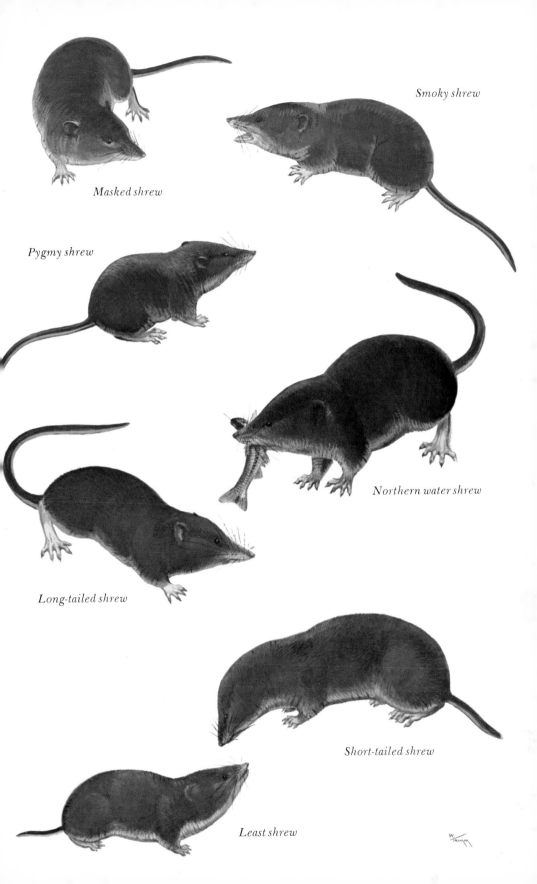

Smoky shrew

Masked shrew

Pygmy shrew

Northern water shrew

Long-tailed shrew

Short-tailed shrew

Least shrew

Northern Water Shrew *(Sorex palustris)*   PLATE 11

The total length is about 5.5″ (head and body about 3″, tail about 2.5″). The upper parts are black or very dark gray, and the underparts are whitish or grayish. The tail is dark above, lighter below. The hind feet are relatively large, slightly webbed between the toes, and with a fringe of stiff hairs along the sides. This shrew lives along the edges of watercourses, lakes, ponds, and marshes. It swims well, and can run across the surface of the water for short distances. It is found throughout most of the Northeast, but is missing from western and southern New York (including Long Island), eastern Massachusetts, Rhode Island, and northern New Jersey.

Smoky Shrew *(Sorex fumeus)*   PLATE 11

This shrew is about 4.5″ long (head and body about 2.75″, tail about 1.75″). The upper parts are dark gray in winter and brownish in summer. Underparts are paler. The tail is brownish above, paler below, and the ears are visible. It is found in moist wooded areas throughout the Northeast, except for Long Island.

Long-tailed Shrew *(Sorex dispar)*   PLATE 11

The total length is about 4.75″ (head and body about 2.5″, tail about 2.25″). Upper parts are blue-gray in winter and dark gray in summer. Under parts are almost the same color as upper parts. The tail of this shrew is considerably longer (relative to the body) than that of the smoky shrew. It prefers rocky areas in moist woods. The Northeastern range is rather spotty, but in general this animal is present in a broad band extending diagonally from northeastern Maine through northern New Hampshire, most of Vermont, western Massachusetts, and eastern and southern New York south to the Catskill region.

Pygmy Shrew *(Microsorex hoyi)*   PLATE 11

The total length is about 3.25″ (head and body about 2″, tail about 1.25″). This is a slender-bodied shrew. Its tail is shorter than in the masked shrew, which it closely resembles. Like the masked shrew, the pygmy shrew weighs less than a dime! The upper parts are brownish, the underparts paler. It lives in open as well as wooded areas, and on rocky, fairly high hillsides. In the Northeast it is found in Maine, the northern two-thirds of Vermont and New Hampshire, and northern and central New York.

Least Shrew *(Cryptotis parva)*   PLATE 11

The total length is about 3.25″ (head and body about 2.5″, tail about 0.75″). The upper parts are brown or gray-brown, the underparts paler. This species looks like a small, brownish version of the short-tailed shrew. It seems to prefer marshes or open grassy areas. In the Northeast it is found only in central, western, and southern New York, extreme western Connecticut, eastern Pennsylvania, and northern New Jersey.

Hairy-tailed Mole *(Parascalops breweri)*   PLATE 10

This mole is about 6.5″ long (head and body about 5.25″, tail about 1.25″). The color is blackish, but slightly paler on the underparts. The tail is short, thick, and covered with hair. The front feet are very broad. The eyes are buried in the fur of the head, and the snout is pointed. It prefers well-drained soils in shrubby fields or woods, and is found throughout the Northeast, except for southeastern Massachusetts, Rhode Island, southern Connecticut, extreme southern New York (including Long Island) and northern New Jersey.

Bats

The Dusk Visitor

On a spring evening one of my greatest pleasures is to conceal myself behind a clump of bushes and look out over the calm waters of our pond. There is always something interesting going on. Perhaps one of the resident muskrats will leave its hole in the bank and cut a wide arrowhead of ripples as it swims across to feed on new, succulent cattail shoots. A gray treefrog will shatter the silence with its clattering trill, and this may stimulate a green frog to gulp its "kung" from among the sedge tussocks. Tree swallows swoop and soar over the surface as they make a final meal of flying insects before retiring for the night. Nearby, a woodcock may utter its sharp "peetz" call as it prepares for its dusk courtship flights. A rabbit appears from nowhere and nibbles at the grass along the top of the earthen dam. It is a peaceful scene.

As the sun drops lower over the ridge I often see a few small bats joining the tree swallows in their hunt for insects or sweeping low over the pond to drink. They are probably little brown bats, but I cannot be certain. Their flight is not as smooth as that of the swallows. They flutter and double back erratically. Sometimes one of them lands on a tree, perhaps to eat a larger than usual insect. More often, after catching an insect in its tail membrane, the bat will swallow it while still in flight.

Several years ago I was watching this activity at the pond when something occurred to make it a really memorable evening. The sun had already set, and shadows were fast gathering. Suddenly, a great bat appeared above the trees at the far end of the pond, and dropped down almost to the level of the dark water. It swept toward me, and as it passed I saw long, narrow wings. It was too dark to make out any other details, but I knew what I was seeing. Of all North American bats, only the hoary bat has a wingspan of fifteen inches or more. In the dusk it seemed twice this size—a truly magnificent bat! It did not stay long. A few passes over the pond and it swept swiftly up over the trees again and was gone, a silent shadow disappearing into the gathering night.

Hoary bats migrate south each fall, many of them apparently going as far as the southern states. Perhaps this particular bat was beating its way back to the forest surrounding some quiet northern lake, there to spend the summer months. I pictured it making its lonely way northward over the dark fields and woodlands, over the twinkling lights of villages and farmhouses. I wondered if it would return to the same location it had left the previous fall—as some birds do—and I wondered how long it would take to get there, how many miles it would cover each night, where it would rest during the day.

I still sit at the pond on spring evenings, and often I think about the great bat that flew there for a few brief moments. Somehow it made a greater impression on me than any of the migrating birds that regularly pass through each year. It is the only hoary bat I have ever been privileged to see at our farm, and I suppose that has something to do with it. But there is something more, a sense of having seen in the gloaming something strangely beautiful, of having experienced something truly awe inspiring.

General Information

Bats constitute the order Chiroptera. Their forelimbs and fingers have become adapted as a framework for membranous wings. These are the only mammals that can truly fly. Although certainly not blind, their vision is usually poor, and they orient themselves and catch their food by means of echo location. While in flight, the bat is constantly emitting extremely high-pitched squeaks. The resulting sound waves bounce back to the bat when they reach a solid object, enabling it to take avoiding action immediately. Thus, bats are able to fly wingtip to wingtip through the pitch darkness of caves without ever bumping into either the cave walls or each other. Because of the importance of picking up these sounds, bats have relatively large ears. While there are fruit-eating, fish-eating, flesh-eating, nectar-lapping, blood-lapping, and insectivorous bats in certain areas of the world, our Northeastern species are all insectivorous. Again, while there are bats with no tails, bats whose tails hang free, and bats whose tails are enclosed by a membrane (the interfemoral membrane), the Northeastern species all fall into the latter category. Among the following bats are some that hibernate and some that migrate to the south with the approach of winter.

Little Brown Bat *(Myotis lucifugus)* PLATE 12

Adult Size. Total length about 3.5″ (head and body about 2″, tail about 1.5″); wingspread about 10″

Description. The upper parts are rich brown, but each hair is black at the base; the underparts are lighter. If pulled forward, the ears do not extend beyond the nostrils. The skin that encloses the tail (the interfemoral membrane) is naked.

Little brown bat catching an insect

Breeding. This bat mates in the fall, but sperm is stored in the female's uterus and fertilization does not take place until the following spring. One young is born in June or July. It is blind and almost without fur, but the eyes open in a few days and the fur grows rapidly. It is weaned at about four weeks, and at that time it begins to fly and hunt for itself.

Habitat. During the day this species roosts in caves, barns, houses, or other sheltered spots away from the light. Just before dark it emerges from its hiding place and can then be found hunting flying insects in almost any location, often in and around villages and farms. It hibernates in caves and other protected sites, but may first migrate several hundred miles south from the more northerly parts of its range.

Food. This animal feeds on flying insects, including a great many mosquitoes. (Most insectivorous bats are said to be able to eat half of their own weight in insects in a single night's feeding!) It uses its interfemoral membrane to scoop insects out of the air, and then removes them with its mouth.

Economic Importance. Insect-eating bats probably do a tremendous amount of good in helping to reduce the populations of mosquitoes and other noxious flying insects. Sometimes bats roost regularly in attics or in the walls of houses. Here their droppings will build up and give off a strong odor. Wire mesh can be fastened over their points of entry to exclude them, or a good application of moth balls scattered around in their roosting site may discourage them from returning.

Comments. One summer I decided to apply a coat of wood preservative to a little wellhouse on my property. I had barely begun when there was a scuffling from within the double walls and one after the other eleven little brown bats wriggled out through a hole in the broken planking and took flight. (Since I wished to be certain of my identification, I caught one of them as it emerged.) Presumably the strong smell of the wood preservative had not agreed with them!

I have often watched what I have assumed to be this species flying rapidly in circles over my pond in the late afternoon. Like other bats, they skim the water to drink. I have waved a stick when one flew near, and quite often it would change direction and make a pass at the stick, without ever touching it. Sometimes, in late summer, both bats and cedar waxwings will hunt flies at the same time over this pond.

The only danger to man involving bats is that some individuals have been known to carry rabies. Many other wild mammals, as well as domestic dogs, are capable of transmitting this disease to man. The incidence of rabid bats is very small, but any bat found on the ground and behaving in a frenzied manner should be treated with caution.

Folklore. There is much legend and folklore regarding bats in general. The old saying "blind as a bat" is completely false. While most bats do not see well, they certainly have obvious eyes and are able to see. Many people believe that bats will try to get into a woman's hair. Again, this is untrue, and in the unlikely event that a panic-stricken bat, trapped in a room, *were* accidentally to become entangled in someone's hair this would not indicate that a catastrophe would occur to that person sometime later—another old belief! It was once thought that the souls of sleeping persons became bats, and that bats should therefore never be molested. Some people think that bats were once birds. Bats are true mammals and have developed completely independently of birds.

As far as vampire bats are concerned, we in the Northeast have nothing to worry about. The only bats to lap (not suck) blood are found in parts of Mexico, south into northern South America. And they are not people who have turned into bats! (The legend of the vampire preceded the discovery of bats that feed on blood, so that the bats were named for the legend, not vice versa.)

Range in the Northeast. Throughout the Northeast; probably the most common species in the region

Similar Northeastern Species. There are four other species that are very difficult to distinguish from the little brown bat. The small-footed myotis is smaller, has dark ears, and its fur is more golden colored. Keen's bat has longer ears (when bent forward they extend beyond the nostril) and less glossy fur. The Indiana bat has dull, gray-brown fur on the back. Each hair is black at the basal half, then grayish, then brown at the tip. The eastern pipistrelle is smaller. Each hair on its back is black at the base, then yellow-brown, then dark brown at the tip.

Silver-haired Bat *(Lasionycteris noctivagans)*   PLATE 12

Adult Size. Total length about 4" (head and body about 2.5", tail about 1.5"); wingspread about 12"

Description. The fur is dark brownish black, with silver tips to the hairs, especially down the center of the back. No other Northeastern bat has this coloring. The ear is almost as broad as it is long, and the interfemoral membrane has fur on its upper side extending about

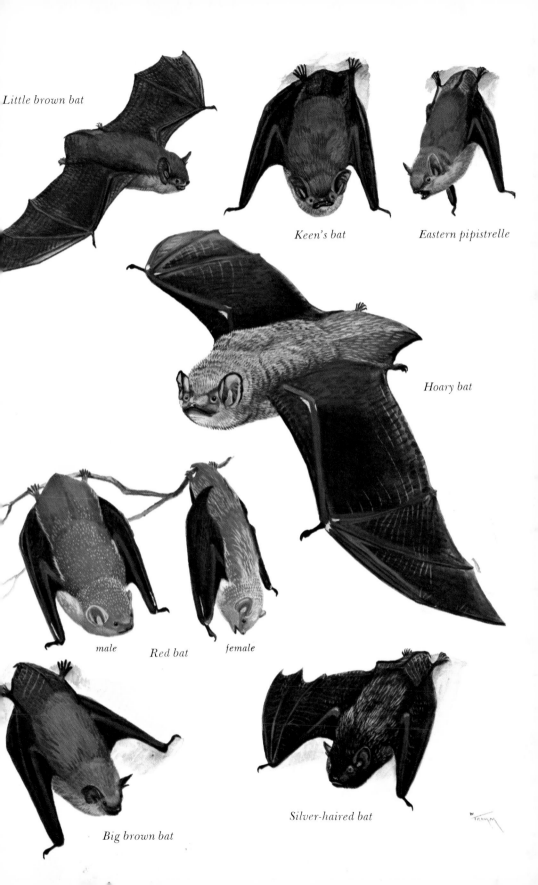

Little brown bat

Keen's bat

Eastern pipistrelle

Hoary bat

male   Red bat   female

Big brown bat

Silver-haired bat

halfway from the base. The flight is quite slow and fairly high. This is mostly a migratory species, although some remain in the north and hibernate while their fellows fly south.

Breeding. One or two young are born between late June and mid-July. After three weeks they are able to fly and hunt for food.

Habitat. This species is seen most commonly in wooded areas where there are streams and ponds, but it may be seen almost anywhere within its range.

Food. It feeds mostly on flying insects.

Economic Importance. See little brown bat.

Folklore. See little brown bat.

Range in the Northeast. Throughout the Northeast

Similar Northeastern Species. The hoary bat is considerably larger and browner, and has silver-tipped hairs over the entire body. Its throat is light yellow-brown, and its interfemoral membrane is completely furred.

Eastern Pipistrelle *(Pipistrellus subflavus)*   PLATE 12

Adult Size. Total length about 3.25″ (head and body about 1.75″, tail about 1.5″); wingspread about 9.5″

Description. Each hair of this small bat is tricolored: dark at the base, yellow-brown for most of its length, and darker brown at the tip. The back appears gray-brown overall. The base of the interfemoral membrane is thinly furred, and the ears are longer than they are wide. Pipistrelles emerge from their roosts fairly early in the evening. They have a weak, erratic flight and can easily be mistaken for large moths. They are mostly hibernators.

Breeding. Usually two young are born between late June and mid-July. For the first few days of life they are carried by the female when she flies, but after this period she leaves them behind. By the time they are three weeks old they are able to fly and to fend for themselves.

Habitat. These bats roost in buildings, rock crevices, caves, etc. They most often hunt in wooded areas near water.

Food. The eastern pipistrelle feeds on flying insects.

Economic Importance. See little brown bat.

Folklore. See little brown bat.

Range in the Northeast. Throughout the Northeast

Similar Northeastern Species. See little brown bat.

Big Brown Bat *(Eptesicus fuscus)*   PLATE 12

Adult Size. Total length about 4.5″ (head and body about 2.75″, tail about 1.75″); wingspread about 12″

Description. The fur of this large bat is long and is dark brown in color. The underside is markedly paler, and the interfemoral membrane is naked. Big brown bats are strong fliers, usually emerging from their roosts rather late in the evening. Some hibernate in buildings or caves; others migrate southward.

Breeding. Usually two young are born in mid-June. They are able to fly and exist on their own at the age of three weeks.

Habitat. This species roosts in buildings, caves, etc. It may be seen flying almost anywhere, including within large cities.

Food. Many beetles are eaten, in addition to other flying insects.

Economic Importance and Folklore. See little brown bat.

Range in the Northeast. Throughout the Northeast

Similar Northeastern Species. The little brown bat and its relatives are much smaller.

Red Bat *(Lasiurus borealis)*   PLATE 12

Adult Size. Total length about 4.5″ (head and body about 2.5″, tail about 2″); wingspread about 12″

Description. The fur is reddish-brown or rust, with a white frosting at the tips of the hairs. Females are browner, with more frosting. The ears are broad and rounded, and the interfemoral membrane is heavily furred on its upper surface. This species migrates in the fall at least as far as southern United States, with some going as far as Bermuda.

Breeding. In mid-June, two to four young are born. They are carried by the female as she flies, but only for the first few days. They are able to exist on their own at four weeks.

Habitat. Red bats often roost in trees or other vegetation, and less often in buildings. One occupied my garage for several months, apparently being able to wriggle in under a loose roof slate. I often disturbed it when I would open the doors during the day, and it would skim my head a few times before seeking the shadows near the roof peak. When hunting, this bat seems to favor wooded areas, but it may also be seen in urban situations.

Food. Its diet consists of flying insects.

Economic Importance and Folklore. See little brown bat.

Range in the Northeast. Throughout the Northeast

Similar Northeastern Species. This is the only Northeastern bat with distinctly reddish-brown fur, and it cannot be confused with any other species in the area.

Hoary Bat *(Lasiurus cinereus)*   PLATE 12

Adult Size. Total length about 5.25″ (head and body about 3″, tail about 2.25″); wingspread about 16″

Description. This is easily the largest of the Northeastern bats. The entire body fur is yellow-brown to very dark brown, with a frosting of silver-tipped hairs. The throat and wing linings are buff. The ears are short and rounded, with naked black rims, and the interfemoral membrane is completely furred. Hoary bats are solitary animals that usually appear late in the evening and that fly high over the treetops in search of food. In the fall most of them migrate to the southern states, but some are known to spend the winter in the Northeast.

Breeding. Usually two young are born, most often in mid-June. As with most bats, they remain with their mother for no more than a month.

Habitat. Hoary bats are most commonly seen in northern, hilly, evergreen forests and along the borders of lakes. They usually roost during the day in trees.

Food. Large moths are eaten, as well as other flying insects.

Economic Importance and Folklore. See little brown bat.

Range in the Northeast. Throughout the Northeast

Similar Northeastern Species. The silver-haired bat is smaller and darker, and its interfemoral membrane is not completely furred on the upper surface. The red bat is smaller and distinctly reddish-brown.

Other Northeastern Bats

These are all closely related to the little brown bat, and are difficult to identify from it, even when held in the hand. They all have a total length of about 3.5″ and a wingspread of about 10″.

Keen's Bat *(Myotis keenii)*   PLATE 12

This bat has less glossy fur than the little brown bat. If bent forward, its ears extend beyond the nostrils. It is found throughout the Northeast.

Indiana Bat *(Myotis sodalis)*

The fur is dull gray-chestnut. Each hair on the back is black on the lower half, then has a narrow band of gray, and then a cinnamon tip. This bat is found throughout the Northeast except for extreme southern New York (including Long Island) and northeastern New Jersey.

Small-footed Bat *(Myotis leibii)*

This is slightly smaller than the little brown bat, with dark ears and gold-tinted fur. There is a black facial mask. It is not widely distributed in the Northeast, and is missing from northern Maine and extreme northern New Hampshire and Vermont.

Carnivores

The Hunters

With the exception of raccoons, which are quite common on our land, the carnivores are rarely to be seen. Sometimes a skunk or a weasel or a red fox will put in an appearance, but such sightings are real events. Once, in our woods, I came upon some newly deposited black bear droppings, but although the bear itself was later spotted by several of our neighbors, I did not see it. Thus, most of my observations of mammalian meat-eaters have been made away from our property. All the more remarkable, then, is the fact that one of the finest views I have ever had of any of them was from our own kitchen windows!

About twenty feet from these windows is a little grassy bank that rises almost to the level of the windows themselves. Early one summer morning, my wife and I saw a gray fox standing on top of this bank and peering in through the windows. It was much bedraggled after walking through the long, dewy grass, and in places its fur was matted with burdock. Its tail hung limp and soaking. But what an alert-looking animal! Its ears were pricked forward, and one paw was raised almost as if it were a bird dog "on point." It was looking from bright sunlight into the shadowy interior of the house, and perhaps that was why it could not see us. After staring into our windows for a moment or two it trotted off around a corner of the house and so out of sight. What a grand start to our day!

There is something about foxes that charms me completely. I remember another encounter with one of these little wild dogs. One warm spring day I was sitting in the shade of a buckthorn tree, near a little path I have cut through the gray dogwood on our hillside. A movement on the open slope above me caught my eye, and I raised my binoculars to see a red fox trotting daintily along just below the skyline.

The fox stopped frequently to sniff at clumps of grass or to nose under rocks. Once it sat and lazily scratched itself with a hind leg. During this process it snapped at the air a few times, where a fly was apparently buzzing around, but the snaps seemed to be more in play than with any serious idea of catching the fly. Its scratching completed, it rose to its feet, contemplated its bushy tail for a moment, and then continued on its way. Suddenly it leaped high into the air, arched its back, and landed with all four feet on a small grass tussock. Probably it had heard or scented a mouse, but this time the mouse was lucky. The fox pushed its nose carefully into the grass once or twice and then gave up. With no further show of interest in searching, it sank to its belly. Apparently it was not really hungry. With its wet tongue hanging from its mouth it lay there in the sunshine and gazed out over the hillside below. I could see its orange-red fur stirring in the breeze, and the occasional twitching of an ear.

After some minutes a broad-winged hawk sailed over the ridge nearby, and the fox looked up at it. The hawk soared out over the valley, and for a few moments the fox watched its flight. Then it stood, stretched itself, yawned mightily, and ambled off over the brow of the hill.

General Information

In the Northeast, the black bear, raccoon, the weasel family, the dog family, and the cat family belong to the order Carnivora. Mammals in this order have teeth that are basically adapted for a diet of flesh, although most of them also eat plant material. All have long, curved, and pointed canine teeth. All have clawed digits. Most Northeastern members of this group live on the ground (are terrestrial), but others live mostly in trees (are arboreal) or are modified for an aquatic way of life. All are important natural controls on rodents and the other small animal life on which they feed.

Black Bear *(Ursus americanus)* PLATE 13

Adult Size. Total length about 5.5', shoulder height about 2.5'; weight normally about 200 lbs. (males are larger, to about 300 lbs.)

Description. This is the only bear present in the Northeast. It is a large, robust mammal with a very short tail, short, fairly rounded ears, and a straight facial profile. The typical color in the Northeast is black, except for the face, which is usually brown. There is sometimes a small whitish patch on the breast. (Elsewhere, especially in western United States, the body color may be almost any shade of brown, and in Alaska there are even some with a bluish or whitish cast.) These bears are mostly nocturnal. They are good climbers and swimmers, and can run for short distances at speeds of more than twenty miles per hour. Their vision is poor, but they have an extremely keen sense of smell. While they do not truly hibernate, they become dormant for long periods in winter, sleeping the time away in caves, windfalls, or other sheltered denning spots. Black bears are normally silent, but they growl or give hoarse roars when angry, and utter a loud "Woof!" when alarmed. Wounded or injured bears often moan or sob very much like humans.

Breeding. Mating usually takes place during the summer. Females mate only in alternate years. In late January or during February, after a gestation period of about 7½ months, the female (sow) gives birth to two young (although sometimes it may be one or three). She bears her young while in her denning quarters. At birth the cubs weigh little more than eight ounces—remarkably small when one considers the size of their mother! They are born blind, and it is almost a month before their eyes open. Cubs grow rapidly and are usually weaned at about 8 months. By the following winter they may weigh over forty pounds. They stay with their mother for considerably more than a year and then are on their own. When about three years old they are sexually mature. Black bears live for about twenty-five years.

Habitat. This is essentially an animal of extensive forested areas, but it may also be found in swampy regions. It is apt to travel widely, however, and occasionally an individual bear will appear on farmland or even in the vicinity of villages.

Food. Although it is a carnivore, this large mammal exists mostly on plant material. It eats much grass and other ground vegetation, and digs industriously for sweet-tasting roots and bulbs. In the fall it feeds on wild berries and fruits, blueberries and blackberries being high on the list of favorites. Most of the meat it eats is in the form of carrion, and here the bear's keen sense of smell plays a major role

Black bear

adult   cub

Fisher

Raccoon

Marten

River otter

in leading it to dead animals or to remains cached by other predators. In spite of its large size, it feeds on mice and other small animals whenever it can find them. Occasionally it will eat a young fawn or a bird's nest containing eggs or young that it may stumble upon, and now and again a particular individual may attack domestic animals. Around campgrounds, the garbage cans and dumps are regularly visited and pawed over. Honey is another favored food.

Economic Importance. The black bear is an important game animal, hunted for both sport and food. The meat is highly esteemed, but must be well-cooked. There have been cases of people contracting trichinosis from eating undercooked bear meat. The fur is of poor quality and is not used in the fur trade. Occasionally black bears kill livestock, but such incidents are very uncommon.

Comments. Wherever this animal has become accustomed to man, such as in campgrounds in the Adirondacks or in northern Maine, it will patrol these areas after dark, overturning garbage cans in order to get at their contents and, in general, creating a ruckus. I have several times had my sleep disturbed by these nocturnal forays and have been glad that I had no food inside my tent to tempt bruin into paying me a closer visit.

This bear is usually afraid of man, but it is potentially very dangerous, particularly if startled or wounded, or if a mother bear feels her cubs are in danger.

During the 1975 hunting season in New York a black bear with an estimated live weight of 750 pounds was taken—a truly enormous black bear!

Range in the Northeast. Maine, New Hampshire, Vermont, western Massachusetts, eastern Pennsylvania, and, in New York, the Adirondacks, Catskills, and Allegheny regions

Similar Northeastern Species. None.

Raccoon *(Procyon lotor)*   PLATE 13

Adult Size. Total length about 32″ (head and body about 24″, tail about 8″); weight 15 to 22 lbs., sometimes heavier

Description. This medium-sized, pointed-faced, bushy-tailed mammal gives the impression of being heavy in the hindquarters. The fur is dense and long. In general, it is yellow-gray in color, with black-tipped hairs overlaid by whitish guard hairs. The face is whitish, with a black mask across the cheeks and eyes. There are about six black rings around the tail, and the feet are white and armed with strong claws.

The raccoon is mostly nocturnal, but is sometimes abroad during the day. It is an adept swimmer and climber. If danger threatens, a mother raccoon will often send her young up into a tree and then try to lead the intruder away. While not a pugnacious animal, a raccoon is able to give a very good account of itself if attacked, and can drive off dogs and other animals much larger than itself.

Breeding. From three to six (usually four) young are born in April, after a gestation period of nine weeks. They have fur, but their eyes do not open for three weeks. By the end of eight weeks they are out hunting for food with their mother, and they stay with her into their first winter. The females often mate before they are a year old. The lifespan is about ten years, but may be considerably longer in captivity.

Habitat. The raccoon is usually found where there are woods and streams, often in swamps or around beaver ponds. It frequently makes its den in a hollow tree, but may also den in rock crevices or even between bales of hay inside a barn.

Food. Almost any available food item is eaten. Raccoons frequently hunt along stream shallows for crayfish and frogs, but feast also on mice, grasshoppers, and birds' eggs and young. Fruits, berries, and nuts are also favored foods, and sweet corn is relished. This animal usually dunks its food if water is nearby. This was once thought to be for the purpose of washing the food (thus the scientific name *lotor*, which means "washer"), but it is now believed that this is merely to make the food easier to swallow.

Economic Importance. The raccoon is of much value in reducing the number of rats, mice, and other small rodents. Its fur was once popular for coats, but today is mostly used for trimming. In the early days of settlement of this country, oil from raccoon fat was often used for treating harnesses and other leather items. When sweet corn is ripe, raccoons will often do a great amount of damage by pulling down numerous stalks and chewing on the ears of corn. They may also ruin berry crops.

Comments. I once found three dead young raccoons that had been run over on a small country road. On stopping to examine them I discovered a fourth young one curled up fast asleep at the edge of the road. Apparently it had not wanted to leave the others. Since it seemed likely to suffer the same fate as its brethren if left there, I grabbed it by the scruff of the neck and carried it, kicking and squealing, up across a field and into some woods. As all who do much country driving will know, raccoons are common highway victims.

One raccoon that we had as part of a college project in wildlife management studies wore a dog harness and was kept chained to a pipe running up the wall. This animal actually learned to unhook the snap fastening to its chain and thus release itself! Although it is illegal to keep a raccoon as a pet, they make charming companions when young—if one can endure their mischievous ways. They can become very affectionate, and quickly adjust to living with people. Unfortunately, they usually become surly and intractable as they get older.

One of the most charming sights I have seen was a family of four young wild raccoons feeding on wild raspberries on a sunny July morning. They would reach up a cane with their little, humanlike hands, bend it over, and then nibble off the fruits.

Range in the Northeast. Throughout the Northeast

Similar Northeastern Species. None.

Marten *(Martes americana)* PLATE 13

Adult Size. Total length about 22″ (head and body about 14.5″, tail about 7.5″); weight about 2 lbs. (females smaller)

Description. The marten is a member of the weasel family and lives mostly in trees. It has soft, shiny fur colored a rich, yellow-brown to fairly dark brown on its upper parts, with the belly and the rest of the underside being paler. The throat and breast are pale orange-buff. Other outstanding features are short legs, rounded ears, a rather foxlike face, and a bushy tail. It is a very inquisitive animal, and has been known to wander into campgrounds apparently just to see what was going on.

Breeding. Although mating takes place during the summer, fetal development is delayed over a long period. The female builds a nest of leaves in a hollow tree or log and gives birth to two to five young in April. Thus, the gestation period is about 8½ months. At birth, the young have a thin coat of yellow hair, but their eyes do not open until they are almost six weeks old. At 3 months they weigh as much as their parents, and they leave their mother for good when about 5 months old. Their lifespan is about fifteen years.

Habitat. This is an animal of the northern evergreen forests where spruce, fir, and hemlock predominate.

Food. While one of its better-known foods consists of the red squirrels that share its habitat, the marten also feeds on mice, birds and their eggs, rabbits, insects, and occasional plant material such as berries and nuts.

Economic Importance. This is a valuable fur-bearer and, unfortunately, is easy to trap because of its inquisitive nature. Overtrapping, combined with lumbering, have greatly reduced its numbers. It is also an important control on rodent populations.

Range in the Northeast. Maine (except for the southwestern part of that state), northern New Hampshire, northern Vermont, and the Adirondack region of New York

Similar Northeastern Species. The fisher is considerably larger, and is very dark brown, with the face and foreparts sprinkled with gray. The mink is more slender, is dark brown, and has a white patch on the chin.

Long-tailed Weasel *(Mustela frenata)* PLATE 14

Adult Size. Total length about 16″ (head and body about 10.5″, tail about 5.5″); weight about 8 ozs. (females smaller)

Description. This mammal has a long, lithe, slender body, a long neck, short, rounded ears, and a long, fairly bushy tail. In summer, the fur is dark brown above and whitish to dark yellow on the underparts, and the tail is brown with a black tip. In winter, most long-tailed weasels grow a white coat (except for the tip of the tail, which remains black), but in southern New York and northern New Jersey this change often does not take place. In their white coats they are known as ermines. While mostly terrestrial, this animal is a good climber and can also swim well (although it prefers not to). It is active both day and night, and is very inquisitive. Its home is usually in the old burrow of another animal or in a rock crevice.

Breeding. Mating takes place in summer but, as with most members of the weasel family, the embryos experience a long period when their development is held back. The young are not born until the following April or May, after a gestation period of about 250 days. There are usually from four to eight young, but sometimes as many as twelve. They are sparsely haired at birth, and their eyes do not open until they are five weeks old, at which time they are also weaned. Females breed at about four months, but males do not mate until much later.

Habitat. These weasels are quite at home in almost any land habitat, but seem to prefer fairly open country.

Food. Most of the diet consists of small mammals such as rats, mice, shrews, chipmunks, and rabbits, but birds, small snakes, frogs, and insects are also eaten. The weasel is apparently one of the very few mammals that kills for killing's sake, not just for food.

Economic Importance. By killing great numbers of small rodents, this animal undoubtedly assists the farmer, but occasionally it becomes a pest by killing poultry. In its winter coat it is in great demand by the fur trade (as ermine).

Range in the Northeast. Throughout the Northeast

Similar Northeastern Species. The mink does not have light underparts, and is more robust. The short-tailed weasel is smaller, more slender, and with a tail only about three inches long. The least weasel is very much smaller.

Mink *(Mustela vison)*   PLATE 14

Adult Size. Total length about 21″ (head and body about 14″, tail about 7″); weight about 2 lbs. (females smaller)

Description. While this is a more robust animal than the long-tailed weasel, it shows a close relationship in having a long, relatively slender body, a long neck, and a long, rather bushy tail. Its color is a rich, dark brown with a white patch on the chin and some white markings on the throat. The fur is soft and shiny. The hind feet are slightly webbed between the toes. This is a mostly nocturnal mammal, but it may be seen frequently by day. It swims very well. The mink has very strong musk—much more powerful than that of the skunk—but is unable to spray it. The musk is not used for defense, but functions as a sexual attractant. When enraged, minks utter hisses and squeals, and may also spit like a cat.

Breeding. Mating occurs in March or April, and the young are born forty-two days later. (Sometimes this gestation period may be as long as ten weeks, owing to delay in the development of the embryos.) The young are born blind, and their eyes do not open until the fifth week. They leave their parents during the summer, and are able to breed before they are a year old.

Habitat. This animal is almost always found near water: along the banks of streams and rivers, in cattail and salt marshes, and along lake shores. The den is usually beneath the roots of a tree on a streambank, but it is also known to inhabit muskrat houses.

Food. The mink feeds on small mammals (including many muskrats), fishes, frogs, crayfish, snakes, and birds. It has been known to store up the corpses of prey animals against a rainy day.

Economic Importance. This is a valuable fur-bearer. Although many minks are raised on farms, the wild mink is still preferred by furriers, for its fur is denser and more lustrous. Its occasional predation on poultry is more than made up for by the many harmful rodents it eats.

Comments. This is a much more common animal than one might suppose; it is found even in urban areas. It is very inquisitive and will sometimes approach people quite closely. One morning a friend and I were eating breakfast while sitting on a large boulder at the edge of a lake in Ontario, Canada. A mink appeared from the nearby undergrowth. It loped up onto our boulder to investigate us, and actually leaped over my outstretched legs before continuing un-hurriedly on its way along the shoreline.

Range in the Northeast. Throughout the Northeast wherever the habitat is suitable

Similar Northeastern Species. This is a much more robust animal than the long-tailed weasel, with which it may sometimes be confused. The only other weasel-shaped animal to be found in an aquatic habitat is the otter, which is much larger and has sleeker fur.

Striped Skunk *(Mephitis mephitis)*   PLATE 14

Adult Size. Total length about 25″ (head and body about 15″, tail about 10″); weight usually about 7 lbs. but may be almost twice as heavy (females smaller)

Description. The skunk is a plump-looking mammal with long fur and a long, bushy tail. The basic color is black, but there is usually much white fur present. A white stripe typically runs from the nose up between the eyes to merge with a large white patch behind the head. From there the patch becomes two stripes that run down over the shoulders and along the sides. The tail is usually black with a white tip. All of these areas of white are quite variable, however, and some striped skunks may be almost completely black.

While they are mostly nocturnal, skunks also may be seen on the move in broad daylight. They are not true hibernators, but they will sleep for long periods in winter whenever the weather is bad.

The skunk is best known for its defensive habit of squirting a strong-smelling liquid at a potential enemy. This liquid is an oil produced by two glands located in the anus. When the skunk is alarmed it will warn of its intentions by first stamping its front feet on the ground and partially raising its tail. If this does not scare away the intruder it then arcs its body around until it is standing in a U shape, with both head and rear pointing at the enemy, lifts its tail high, and very accurately fires a stream of oil at the eyes. This stream may carry for a distance of more than fifteen feet, and causes intense burning pain if it strikes the eyes. (Water will quickly remove the burning sensation.)

Breeding. Mating takes place during February or March. After a gestation period of about eight weeks, from four to eight blind young are born. Their eyes open at about three weeks, and they are weaned at eight weeks. The mother takes her family out hunting once the young are able to walk well, and there are few more charming sights than seeing the little skunks following their mother in single file. They usually leave her for good when fall arrives.

Female striped skunk and young

Habitat. Although it is a creature of fairly open countryside, including farmland, the skunk is often seen in well-wooded areas. The nest, of dry leaves and grasses, is constructed in an old fox or woodchuck den, or in a burrow dug by the skunk itself, or in a rock pile or other protected site above ground.

Food. This is the most omnivorous of the Northeastern members of the weasel family, feeding on mice, chipmunks, insects (including honey bees), berries and fruits, carrion, and garbage.

Economic Importance. The skunk has been known to kill poultry, but it also destroys many mice and other small rodents, and is, therefore, mostly beneficial as far as man is concerned. The fur is used mostly as trimmings for coats, but sometimes entire coats are made from the pelts. Oil from the anal glands has been refined and used as a base for perfume.

Comments. This is a completely inoffensive animal, and will discharge its oil only as a last resort. Gasoline is effective in ridding oneself of the smell, but must be washed off at once to avoid burning the skin. It is said that ketchup or tomato juice can be used to remove the odor from a "squirted" dog.

"Squirting" stance of the skunk

Striped skunk

Mink

Long-tailed weasel

winter

summer

summer

Short-tailed weasel

winter

summer

winter

Least weasel

It seems to be widely believed that if a skunk is gripped by the tail, with its feet off the ground, it will be unable to squirt. DON'T TRY IT!

Range in the Northeast. Throughout the Northeast

Similar Northeastern Species. None.

Red Fox *(Vulpes vulpes)* PLATE 15

Adult Size. Total length about 39″ (head and body about 24″, tail about 15″); weight about 10 lbs. (females smaller)

Description. This long-haired, pointed-nosed, large-eared, and bushy-tailed mammal looks very much like a fairly small dog. The normal coloring is a golden yellow-red above and white on the underside. The legs, feet, and the rear of the ears are black, and the tip of the tail is white. There are other color phases, however, and a litter of pups may include a black fox with a frosting of white over the back (the "silver fox"), a yellow-brown animal with a dark band over the shoulders crossing another one running down the back (the "cross fox"), or many other intermediate phases. Nevertheless, all are red foxes, and all have a white tip to the tail. This species is mostly nocturnal, but is often active by day. While normally proceeding at a brisk trot, this small member of the dog family can run for distances of up to a mile at a speed of more than twenty-five miles per hour. It swims well. These foxes have a variety of quiet calls for communicating with their young, but the loudest and best-known sound is the hoarse bark uttered by the males during the mating season in late winter.

Breeding. Mating takes place in late winter, and after a gestation period of fifty-one days, from four to nine young are born with their eyes not yet open. The eyes open after a week, but the pups do not emerge from the den until they are about five weeks old. For a short period after the female (vixen) has given birth, the male brings food to the den for her. After the pups have been weaned, both the male and the female go out hunting food for the young. Once the pups are old enough to leave the den, they spend much time in playing together, to prepare their muscles for hunting when they are on their own. At this stage they are among the most engaging of all wild creatures, interested in everything and often quite unafraid. I was once filming four red fox pups playing at the entrance to their den. The whirring sound of the camera proved irresistible to one of them, and the plump little fellow dashed down from the den entrance, peered up at the camera from between the legs of the tripod, and then turned about and raced back to the others. Pups remain with their parents until the fall and breed the following year.

Habitat. Although tending to be an animal of the open countryside (farmland, marshes, open hillsides, etc.), the red fox is also to be seen in thinly wooded areas, and may even be found within a heavily forested tract. The den can be located out on an open hillside or well back in the deep woods. It is either dug by the fox itself or may be merely an enlarged woodchuck burrow. If a fox den is occupied, there is usually a pronounced "gamey" odor hanging around the entrance holes.

Food. Depending upon the season, food items consist of mice, woodchucks, rabbits, ground-nesting birds and their eggs and young, turtles and their eggs, frogs, insects, grass, berries and fruits, and carrion.

Economic Importance. Although the red fox sometimes kills poultry, it is very beneficial in consuming great numbers of rats and mice. It is an important fur-bearer, but is also hunted for "sport" by horsemen and hounds. This barbaric activity is especially popular in England and Europe, but is also carried out in some parts of the United States. This animal seems to be particularly susceptible to rabies. Although often contracted by contact with a rabid domestic dog, the disease spreads rapidly in red fox populations, and there have been serious epidemics in the Northeast. Another, sometimes epidemic, red fox killer is a form of mange caused by mite infestation.

Range in the Northeast. Throughout the Northeast, even close to large urban centers

Similar Northeastern Species. Its closest relative, the gray fox, does not have black feet or a white tip to the tail; nor does the coyote, which is larger and grayish. The fisher is very dark colored and has short ears, and the marten is much smaller, with a buff patch on the breast. (The white tail tip is also lacking in these two species.)

Gray Fox *(Urocyon cinereoargenteus)*   PLATE 15

Adult Size. Total length about 38″ (head and body about 24″, tail about 14″); weight about 9 lbs., but sometimes more

Description. The gray fox is shaped much like the red fox, but its muzzle and legs are shorter. The upper parts are mostly frosted gray or blackish, but the back of the ears, the sides of the belly, and the

sides of the neck and legs are rusty brown. The belly is white. The long tail has a dark stripe running down its upper surface to the black tip.

Like the red fox, this is a mostly nocturnal animal, but unlike the red fox it is capable of climbing trees. It may do so to escape enemies, or sometimes just to rest in safety.

Breeding. Breeding habits are very much like those of the red fox, but pups may be born on into May after a slightly longer gestation period.

Habitat. In the Northeast it prefers wooded areas, although it may sometimes be seen in more open country.

Food. The diet of the gray fox is similar to that of the red fox, but it may eat a greater variety of plant foods.

Economic Importance. This fox kills very little poultry, but is a good control on many small animal pests, particularly mice and rabbits. Its fur is used for trimming, though it is not valued as much as that

Gray fox in a tree

of the red fox.

Range in the Northeast. Throughout the Northeast, but only in the western part of Maine; rare on Long Island, now only at the eastern end

Similar Northeastern Species. The red fox has a white tip to the tail, regardless of its other coloring. The coyote is larger and grayish, with no "pepper and salt" markings and without the well-defined rusty coloring. The fisher is very dark colored and has short ears, and the marten is smaller, with a buff patch on the breast.

Bobcat *(Lynx rufus)* PLATE 15

Adult Size. Total length about 33″ (head and body about 28″, tail about 5″); weight about 20 lbs., but sometimes to more than 30 lbs.

Description. The outstanding feature of this cat is the short tail, the tip of which is black above and white below. The fur is quite short and dense, and (in the Northeast) is usually chestnut brown marked with black spots and streaks. In winter it is grayer. The underside is white with dark, scattered spotting. The ears are pointed, with a short tuft of black hairs at the tips and with the rear of the ears being black with a white spot in the center. In remote areas this animal sometimes may be seen during the day, but it is essentially nocturnal. It is a good climber, but usually stays on the ground. Likewise, it swims well, but prefers not to enter the water.

Coyote

Gray fox

Red fox

Lynx

Bobcat

Although normally silent, the bobcat screams and yowls loudly during the mating season. When angry it snarls, growls, hisses, and spits.

Breeding. Mating is usually in late February, and the one to four young are born about fifty days later. The kittens have fur at birth, but their eyes do not open for about ten days. They feed on their mother's milk for about eight weeks and then begin taking solid foods. They leave their mother in the fall. The den is in a rock crevice, a windfall, or some other sheltered spot. However, the bobcat is quite a traveler, and may stay away from its home den for several days at a time.

Habitat. This animal lives mostly in woods, hills, and swampy areas, but is sometimes seen on open farmland.

Food. Bobcats feed mostly on small mammals and ground-nesting birds, but they are quite capable of killing foxes, porcupines, and even deer if the deer is bogged down in deep snow. They will eat carrion if fresh meat is not available. I once saw one of these animals nibbling at grass, but such plants are probably pastime foods rather than part of their nutritional diet.

Economic Importance. The fur is used for coats and trim, but, while quite attractive-looking, is not very durable. Although individuals occasionally steal chickens—and may kill sheep and lambs—they do this only rarely, and are certainly of value in feeding on a variety of rodents and other animals that are often pests.

Range in the Northeast. Every Northeastern state except Rhode Island, but absent from eastern Massachusetts, eastern and central Connecticut, extreme southern New York (including Long Island) and northeastern New Jersey

Similar Northeastern Species. The lynx is grayer, has longer fur and longer ear tufts, and the tip of the tail is completely black.

Other Northeastern Carnivores

Fisher (*Martes pennanti*)   PLATE 13

The total length is about 38″ (head and body about 24″, tail about 14″), and it weighs about 10 pounds. This is a large, fox-sized member of the weasel family with short, rounded ears, a bushy tail, and thick legs and neck. The fur is a very dark brown, with scattered, white-tipped hairs, especially on the face and shoulders. It is found in forested areas (including evergreen forests) or sometimes in swamps. The fisher takes readily to trees in search of food, and will chase red squirrels and martens through the branches. Other prey consists of most small mammals, foxes, and even deer (especially if the deer is bogged down in deep snow). It also kills birds such as grouse, and may feed on berries and other plant material at times. It can swim well, but spends little time in the water—in spite of its name. The fisher is an important fur-bearer that today seems to be slowly reoccupying areas from which it had been eliminated by over-trapping. In the Northeast it is normally found only in Maine (except southwestern Maine), the northern half of New Hampshire and Vermont, and the Adirondack Mountains of New York.

Short-tailed Weasel *(Mustela erminea)*   PLATE 14

This animal is about 11″ long (head and body about 8.25″, tail about 2.75″) and weighs about 3 ounces. Its size is barely larger than that of a chipmunk. The short-tailed weasel is colored much the same as the long-tailed weasel and is most common in evergreen forests (but may also be seen in farming areas, often along old stone walls). It feeds on mice, insects, small birds, and occasionally rabbits. Because of its small size, this weasel is not as important to the fur trade as the long-tailed weasel. It ranges throughout the Northeast.

Least Weasel *(Mustela nivalis)*   PLATE 14

The adult is about 7″ long (head and body about 6″, tail about 1″), and it weighs about 1.5 ounces. This is the smallest member of the carnivores in all of North America. In summer its upper parts are a rich brown, with white underparts and toes. The tip of the tail has some black hairs in it, but is not completely black. In the more northerly parts of its range this creature may become white in winter (except for the black hairs at its tail tip). Least weasels live in heavy forests, but may sometimes be seen in more open country. They probably feed mostly on mice. In the Northeast they are found only in western New York and eastern Pennsylvania.

River Otter *(Lutra canadensis)*   PLATE 13

The total length is about 40″ (head and body about 26.5″, tail about 13.5″), the weight about 12 pounds. This aquatic member of the weasel family has a broad, flat head, small eyes and ears, webbed toes, and a heavy tail that tapers toward the tip. The fur is short and dense, dark brown above and lighter below. It is found where there are lakes and large streams, but it is seldom seen because of its extreme shyness. Otters are expert swimmers, underwater as well as at the surface. They apparently enjoy playing, and are known to slide repeatedly on their bellies down clay or snow-covered banks. They feed on fishes, crayfish, aquatic insects, amphibians, and other small animal life that they catch in or near the water. The fur is lustrous and very durable, and therefore is valuable to the fur trade. Otters live in suitable areas throughout much of the Northeast, but are absent from western Connecticut, southern New York (south of the Catskills), eastern Pennsylvania, and northern New Jersey.

Coyote *(Canis latrans)* PLATE 15

The coyote is about 48″ long (head and body about 34″, tail about 14″). The weight averages about 30 pounds but varies from about 20 pounds to over 40 pounds. It is a wolf-shaped animal with longish fur, large pointed ears, and a bushy tail. On the back and sides the overall color is mostly gray or brown-gray, with black tips to many of the hairs. The legs and feet are tinged with reddish, and the underside is white. In the Northeast, coyotes inhabit wooded areas and brushy farmland. The voice is a high-pitched yapping. These animals are very efficient rodent-catchers and certainly, as a species, are far more beneficial than detrimental. Although this species is thought to have occurred in the Northeast only during the past few decades (possibly as a result of animals that originally escaped from captivity), there are now fairly frequent sightings from various locations in Maine, Vermont, New Hampshire, western Massachusetts, and New York.

Lynx *(Lynx lynx)* PLATE 15

The adult is about 34″ long (head and body about 30″, tail about 4″) and usually weighs 15–30 pounds, but sometimes more. At first glance, this cat may be confused with the bobcat. The lynx differs, though, in that the tip of its tail is completely black (rather than black above and white below) and in having longer legs, larger feet, and more prominent ear tufts. There is a large ruff of hair surrounding the face, and the eyes are large and yellow. The color of the fur is usually a soft gray with pale brown speckling. The lynx is mostly nocturnal, and it is a good climber and swimmer. Although, when they are available, snowshoe hares seem to be its favorite food, the lynx also feeds on squirrels, mice, grouse, foxes, and some carrion. It is sometimes able to kill deer when they are hampered by deep snow. The lynx is found mostly in coniferous forests, but in the Northeast is present only in the more northern parts of Maine, New Hampshire, Vermont, and New York. Even in these areas it is rare.

Mountain Lion *(Felis concolor)*   PLATE 10

The adult may be up to about 8′ long (head and body about 5.5′, tail about 2.5′). It weighs about 150 pounds. Females are smaller. This very large cat has a relatively small head with rounded ears, and a *long,* cylindrical tail. The fur is short, colored light brown or grayish above and lighter below. There is a dark tip to the tail. The mountain lion (also called cougar, puma, panther and painter) prefers wooded country where there are gullies and rocky hillsides. Although it feeds on large mammals such as deer, it also eats birds and almost all of the smaller mammal species. It is normally afraid of man, and gives him a wide berth. Most "mountain lion" sightings are actually of bobcats. (People forget to look for the long tail.) Mountain lions, however, were once widespread in the Northeast. They are still present across the border, in New Brunswick, and it is possible—even likely—that they will spread back into Maine and some of the other Northeastern states.

Rodents

A Balancing Act

Rodents include some remarkably interesting mammals. Among these may be numbered the chipmunk, a pert, precocious, pint-sized squirrel known to most people living in rural areas. I could tell many stories about chipmunks, but my favorite concerns one that lived in our garden for several years —one whom we named Charlie. His tail made him easily recognizable, for at some point in his life it had been broken. When the break knitted it left the tail bent at an acute angle in its center.

On an afternoon in October I picked a large basketful of apples and carried it home. There I deposited the basket on the garage floor, near the door. Ten minutes later I was in the room adjoining the garage when I heard a slight crunching sound. I looked into the garage and there beheld Charlie, standing on his hind legs and chewing away at an apple that showed through a hole in the basket. The apple was now somewhat disfigured, so I decided to donate it to him. I removed it from the basket, set it on the floor and retired around the corner to watch events. The chipmunk, who had fled in panic at my sudden appearance, shortly returned. He sneaked silently through the doorway like a little striped shadow, noticed the apple, and at once moved to it.

Now began a real struggle, for he was determined to make off with the whole thing. He dug his teeth into it and several times tried to walk off with it held in front of him. But the apple was too large; too heavy for his teeth to hold. He rested for a moment or two. Then he fastened his teeth into it again and with a sudden heave raised it aloft over his head. Looking somewhat like a sea lion balancing a large rubber ball on its snout, he marched on tottery legs to the door and staggered off into the sunlight.

Thus far all had gone well. But now he had to negotiate the route to his burrow, and along this route lay all manner of obstacles. With his vision blocked by the apple he hit a good 90 percent of them! He walked into rocks, he was knocked over by poplar seedlings, he became entangled in weeds and shrubbery, but he persisted. Each time he collided with something and dropped the apple he doggedly raised it over his head again and, stiff-legged, marched on his way until he hit another obstacle. In this fashion he gradually progressed along the base of the garage wall and disappeared around its corner. I followed, and peered around the corner.

Charlie had arrived at his burrow entrance, and was trying to push the apple into the hole. But the apple was much larger in diameter than the opening, and his efforts were in vain. After much energetic shoving he paused to rest. He next tried backing into his burrow and dragging the apple in after him. This was even more frustrating, and he soon gave that up. His next move was to chew the apple into two pieces, and to then take each piece down into his home separately!

Now, assuming that he had done this many times before, I suppose one could classify this as learned behavior. But this would be quite an assumption, for there are no apple trees anywhere near the home range of this particular animal. To me, the behavior of this chipmunk seems to show glimmerings of an ability to solve this particular problem—albeit by trial and error.

Any damage our chipmunks may do in the garden is far outweighed by their value as entertainers. Long may they flourish!

General Information

On a worldwide basis, the order Rodentia includes more species and more individuals than all other orders of mammals combined. Northeastern rodents are made up of squirrels and their relatives, rats and mice, jumping mice, the beaver, and the porcupine. They are best known for their gnawing activities, accomplished by a pair of incisor teeth in both upper and lower jaws. These teeth grow from the roots throughout the lifetime of the animal, but are kept to a constant length by rubbing against each other at the point where the tips of the upper incisors meet those of the lower ones. There are no canine teeth, so that there is a large gap between the incisor teeth in the front of the mouth and the grinding teeth further back in the cheeks. In this order are to be found the best nest-builders and the most efficient food-hoarders among all of the mammals. While many rodents can be pests as far as humans are concerned, they may also assist man by aerating the soil with their tunnels and runways, reforesting barren areas, eating weed seeds, providing fur, and being used experimentally in combating human disease. They also function as food for a great many other kinds of mammals, birds, and reptiles, thus reducing the pressure that these wild predators might exert on domestic livestock.

Woodchuck (Groundhog) *(Marmota monax)* PLATE 10

Adult Size. Total length about 24″ (head and body about 18″, tail about 6″); weight about 8 lbs.

Description. This large, stocky-bodied rodent has short, rounded ears, short legs, and a rather long, bushy tail. Its fur is basically brown, with a fairly heavy frosting of white. The cheeks are dull white. The woodchuck is mainly diurnal, feeding most actively in the early morning, late afternoon, and early evening.

Breeding. Males and females usually come together only during the breeding season in March or early April. About a month later, two to five blind, naked young are born. They remain in the den until they are one month old, by which time their eyes are open and their fur has grown. Then they emerge and begin fending for themselves. By midsummer they have left the den and moved off to dig their own burrows.

Woodchuck on hind legs

Habitat. Woodchucks tend to prefer meadowland, open fields, and brushy hillsides, but I have often seen them well within thickly wooded areas. Along a forty-mile stretch of the Taconic Parkway, in New York, I have several times counted more than fifty feeding at the grass road verges. The den is an extensive burrow system with one main entrance and several well-hidden "pop holes" for emergency use. The main entrance is larger than the others and has a mound of excavated soil and stones at its opening. Below ground are tunnels and chambers that may run for thirty or forty feet and that may go as deep as six feet. Some of the chambers are used for sleeping, and may have snug nests of grasses and other plants in them; others may be used as lavatories.

Food. While the woodchuck may eat some insects, most of its food is made up of grasses, wildflowers, thistles, berries, fruits, and other plant material.

Economic Importance. Although consuming many weeds, and so helping in their control, this mammal unfortunately is also fond of agricultural crops and garden vegetables. It can do a great amount of damage in a short time, especially in gardens. Until I fenced in my garden—with a strand of electrified wire running four inches above the top of the fence—woodchucks would consistently eat entire rows of young lettuce and other tasty crops. (In putting up an anti-woodchuck fence, remember that this is a burrowing animal! Bury a good eight inches of your wire fence below the ground, otherwise a woodchuck will dig right under the fence.)

Stewed or roasted woodchuck is quite tasty, but the meat is coarse, and must be soaked for at least half a day before cooking.

Comments. The woodchuck is a true hibernator. During the late summer it gorges itself on plants until it is very fat. In early October it retires to its winter quarters to begin its long winter sleep. (Although in New York I have sometimes seen woodchucks still up and about in mid-November.) Even though it may have a den out in the fields during the summer, it usually has its winter burrow within a wooded section. Here it curls into a tight ball and stays dormant until early March. During this period its heartbeat slows down, its respiration is retarded, and its temperature drops. Its reserve of stored fat is used up so slowly that it is sufficient to sustain the woodchuck throughout this entire period.

Like any rodent, the occasional woodchuck suffers from what dentists call malocclusion—the upper incisor teeth fail to meet the lower ones. When this happens the animal is usually doomed, for the teeth continue to grow from the roots and eventually prevent it from feeding. One such animal I found had one of its lower incisors curving right out of the mouth and back up into the nostril, while the upper incisor curved downward like the canine tooth of a saber-tooth tiger.

Although one does not think of this animal as being arboreal, I have often seen one scrambling around in small, wild cherry trees in the early spring, feeding on tender leaf buds. When I have been spotted by the woodchuck it will plummet to the ground—sometimes eight feet or more—without any apparent injury.

The voice is a short, fairly low-pitched whistle. When close enough, I have also heard a rather quiet, quavering, and quite musical whistling that gradually fades away. If it is angry, this animal growls and grinds its teeth.

This is an extremely cautious animal, rarely feeding for more than half a minute or so without stopping to look around. Often it rises onto its hind legs in order to survey the landscape for potential enemies such as foxes, domestic dogs, or man.

Folklore. As most people know, 2 February is "Groundhog Day," when the woodchuck supposedly emerges from its burrow for the first time each year. If it does not see its shadow it knows that winter is over, but if it casts a shadow it knows that there is more bad weather to come, and it goes back to sleep for another six weeks. Needless to say, this is merely a pleasant little story and has absolutely no basis in fact.

Range in the Northeast. Throughout the Northeast

Similar Northeastern Species. The muskrat is smaller and has a long, naked tail. The beaver is larger and has a broad, horizontally flattened tail.

Eastern Chipmunk *(Tamias striatus)*   PLATE 16

Adult Size. Total length about 9.5″ (head and body about 6″, tail about 3.5″); weight about 3 ozs.

Description. The chipmunk is a small member of the squirrel family. It has fairly large, rounded ears, short legs, and a relatively large, long-haired but flattened tail. The tail is carried vertically when running. The fur on the upper parts is a grayish or reddish brown with a sprinkling of lighter hairs and with a bright chestnut rump. The outstanding feature is the striped pattern on the back. There are five dark stripes running from the shoulder area to the rear of the body (one down the center of the back and two on each side). The two that run along each side have a buff-white stripe between them. There is also striping on the head: a buff stripe above and beneath each eye, and a short, dark stripe running from the rear of the eye. The general effect is of a broad white eye-ring. The tail is darker than the general color of the upper parts, and the underside of the body is white.

There are large, internal cheek pouches. As it gathers its supplies of seeds, these cheek pouches bulge more and more, giving the little animal a ludicrous appearance. When the pouches are stuffed to capacity it retires to empty them out in its burrow, after which it returns for another load! These pouches are quite possibly also used for carrying away soil excavated from the burrow.

The alarm note is a loud, sharp, high-pitched "chip!" almost bird-like in quality. Other calls sound more like chirrups—for example, a less strident "chick-chick-chick" that may continue uninterruptedly for several minutes. In order to look for enemies such as hawks, weasels, foxes, and housecats, the chipmunk will often stand upright, balancing itself with its tail, and peer suspiciously around before going back to the serious business of food gathering.

Breeding. Mating is in early April, and after a gestation period of thirty-one days two to five tiny blind, hairless, and pink-skinned babies are born. In about two weeks they have a fine coating of hair, but it is a month before their eyes open. They remain with the female for about three months. Although young chipmunks are sometimes seen in late summer, it is thought that, rather than being from a second litter, they may be young from a late-breeding female.

Habitat. Chipmunks are apt to be found in almost any reasonably dry area (but not open fields): forests, woodlots, brushy hillsides, farm yards, stone walls, suburban gardens and lawns, and even city parks. The home is a long, winding tunnel about two inches in diameter that may extend for more than thirty feet. From the entrance, the tunnel drops vertically for a few inches and then slants down for another two or three feet. Several chambers open out from the tunnel; some of these are used for storing food, one for sleeping, and one as a toilet. The sleeping chamber contains a warm nest of leaves and grasses, with an emergency supply of food beneath it. Excavated soil seems to be carried away from the immediate vicinity of the burrow, for certainly there is never any soil to be seen near the entrance hole.

Food. The main diet consists of a wide variety of seeds, nuts, berries, bulbs, and fruits, but chipmunks also eat many insects, slugs, snails, and other small animal life. They do not disdain birds' eggs and meat. The former are sometimes stolen from poorly concealed nests, and the latter consists of an occasional mouse. During the warmer parts of the year there are usually a couple of chipmunks picking up seeds from the ground below my bird feeders, in company with a variety of birds. I have never seen one attack a bird, but once, when a young robin smashed into my kitchen window and broke its neck, a chipmunk immediately leaped upon it and proceeded to eat a good half of it before returning to the seeds on the ground.

Feeding takes place at odd intervals during the day, from soon after dawn until early dusk.

Economic Importance. Chipmunks can damage lawns and flower beds with their burrows, and will sometimes feed on the bulbs of garden plants, but they make up for this by eating many weed seeds and small animal garden pests. They should be denied entrance to a house, for they can wreak much havoc in a short time by trying to gnaw their way out!

Comments. While it may hibernate during long cold spells, this little animal awakens periodically, and on fine days in midwinter I have seen its tracks in the snow. Its underground food store enables it to feed well when food is unavailable outside.

Although not as much at home in trees as the true tree squirrels, chipmunks are expert climbers. My bird feeder stands on a six-foot painted metal pole, and the paint was enough to give chipmunks a grip, for they would regularly run up the pole and onto the feeder. (I solved this problem by nailing a two-pound coffee can upside down under the feeder, with the pole running through it!)

Periodically there seem to be population peaks when these amusing little creatures are extremely abundant. At such times my lawns and gardens have been pockmarked with the entrance holes to their burrows, and there have been dozens of them around the house and its environs, fighting with each other when territories have been invaded, squeaking angrily, and in general creating quite a disturbance. I have, therefore, had to take drastic steps to reduce their numbers. This has involved live-trapping them (they are very easily baited into live-traps), loading them into the car, driving off for a distance of five miles or so, and then releasing them into wooded areas. During one such chipmunk population explosion I trapped twenty-four from right around the house (in two days)—and there were still quite a few left when I had finished!

Range in the Northeast. Throughout the Northeast

Similar Northeastern Species. None. This is the only striped member of the squirrel family in the Northeast.

Eastern Gray Squirrel *(Sciurus carolinensis)* PLATE 16

Adult Size. Total length about 17″ (head and body about 9″, tail about 8″); weight about 1 lb.

Description. This is a large, robust tree squirrel with rounded ears and a long, bushy tail. The upper parts are grayish, but with a light brown wash on the face and, in summer, on the sides and legs. The underside is white. The tail is darker at its base and is frosted with silver.

The gray squirrel is a diurnal animal, especially active in the early morning and late afternoon throughout the year. When alarmed, it flattens itself against the trunk of the tree and moves constantly around the trunk to keep itself always shielded from danger. It swims

Southern flying squirrel

Northern flying squirrel

Red squirrel

Fox squirrel

Eastern gray squirrel

Eastern chipmunk

well. The call is a sharp, barking "ka-ka-ka," and often its tail is flicked in time with each note.

Breeding. The home is located in a tree cavity or consists of a bulky nest of leaves and twigs built high up among the branches. Most young are born during March or April, but a second litter may be produced during the summer. The gestation period is forty-four days. There are usually two to four young, and they are blind and hairless at birth. The eyes do not open for about five weeks, and it is eight weeks before the young squirrels are active outside of the nest. Although they may breed at twelve months, they are not fully grown until they are two years old.

Habitat. Gray squirrels are found mostly in deciduous woodland, especially among mixed hardwoods where oaks and hickories predominate. This is a species that has become accustomed to living close to man, and that is now very common in city parks and suburban gardens.

Leaf nest of the gray squirrel

Food. Acorns and other nuts are eaten, and also seeds, berries, buds, insects, birds' eggs, and, sometimes, nestlings.

Economic Importance. While it is of some value in eating harmful insects and weed seeds, this squirrel sometimes becomes a pest in vegetable gardens or at bird feeders. Its flesh is edible and is enjoyed by some people. It has a strong "gamey" flavor.

Comments. "Melanism," an overabundance of a black pigment called melanin, occurs quite frequently in this squirrel. Melanistic gray squirrels may be quite black in color or may have brown hairs interspersed with black hairs. In some areas there are quite large numbers of these handsome creatures.

Range in the Northeast. Throughout the Northeast except for northern Maine

Similar Northeastern Species. The fox squirrel is larger, and usually has a black head and white ears and nose. It is found only in extreme western New York. The red squirrel is quite a bit smaller, reddish or olive-gray in color, and has a white eye-ring.

Red Squirrel *(Tamiasciurus hudsonicus)*   PLATE 16

Adult Size. Total length about 13" (head and body about 8", tail about 5"); weight about 7 ozs.

Description. The reddish color and small size of this stocky tree squirrel at once identify it. In summer the fur tends to be rather olive-yellow above, and there is a black line dividing the upper parts from the white underside. In winter the back is reddish, the sides are grayish, and tufts of reddish hair grow from the tips of the ears. There is a white eye-ring.

Certainly one of the outstanding characteristics of the red squirrel is its voice. It is very jealous of its territory, which usually consists of an area of about two or three acres, and it is not slow to vent its feelings upon any trespasser. Its loud scolding is made up of a whole repertoire of sounds. One that lives in my pine woods will come rushing out through the branches at an amazing speed, sputtering with resentment, and take up a position on a branch over my head. Here it goes into a frenzy of explosive churrings, barks, squeals, squeaks, spits, and growls, stamping its feet and jerking its tail in rage. Since I move away after watching this performance for a few minutes, it must think its violent display of temper scares me off! Long after it is out of my sight I can still hear it muttering and grumbling to itself, and occasionally uttering a strident "Wuk!"

Like the gray squirrel, this animal is active during the daylight hours all year long.

Breeding. Mating takes place in March or early April, and after a gestation period of about forty-two days a litter of (usually) three to six young is born. The baby squirrels are blind and hairless. They acquire a light coat of fur by the tenth day, but their eyes do not open for four weeks. It may be a further month before they are weaned. They leave their mother during the late summer, at which time she may give birth to another litter.

Red squirrel's cache of drying mushrooms

Habitat. Red squirrels seem to prefer areas where conifers such as spruce, pine, and hemlock grow, although they may be seen in deciduous woods. The nest is made of grasses, bark shreds, and leaves. It is usually in a tree cavity, but may be a leaf nest in a limb crotch close to the trunk, or even a burrow underground, often beneath a stone wall. For several years I had one of these squirrels living between the double walls of an old wooden wellhouse. I often saw this little fellow sunning itself on the roof of the wellhouse or heard it scrambling around between the walls when I chanced to disturb it.

Food. In addition to nuts, seeds, berries, tree buds, and flowers, this squirrel is very fond of mushrooms and other fungi. It also eats birds' eggs and nestlings and some insects. In coniferous forests the red squirrels make sizeable heaps of several bushels of cones. These heaps are visited regularly, and are usually surrounded by the husks of cones that have been ripped apart so the seeds can be eaten.

Economic Importance. The fur is sometimes used in trimming coats.

Folklore. In the United States there is a common belief that the red squirrel will chase and castrate any male gray squirrel it finds.

Range in the Northeast. Throughout the Northeast, but not on Long Island

Similar Northeastern Species. The gray squirrel is much larger, grayish, and not nearly as vociferous!

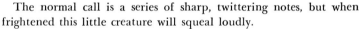

Adult Size. Total length about 10.5″ (head and body about 5.5″, tail about 5″); weight about 5 ozs.

Description. This small animal has a loose fold of skin stretching along each side from wrist to ankle, a wide, furry, very much flattened tail, and large, shining eyes. The ears are large and quite rounded. The fur is very dense and soft. In winter it is colored red-brown on the upper parts and white on the underside (but with each belly hair having a dark base). In summer the fur is darker on the back, and the underparts are more buff.

Although named "flying" squirrel, this is actually a glider. When it wishes to move from one tree to another it climbs high up on the trunk and then leaps out into the air, spread-eagling its arms and legs. The membrane between its wrist and ankle is thus extended and acts somewhat like a kite as the squirrel glides downward to the next tree. Just before reaching its target the squirrel jerks its tail upright and loops upward to land. It then runs up the trunk and repeats the process to the next tree. Some of these glides may be of more than fifty yards, and by twisting its tail and lateral membranes it steers itself quite efficiently through the air. Because it is entirely nocturnal this fairly common animal is rarely seen.

Flying squirrel

The normal call is a series of sharp, twittering notes, but when frightened this little creature will squeal loudly.

Although not a true hibernator, the flying squirrel will become dormant for long periods during especially cold weather.

Breeding. After a gestation period of forty days, three or four young (sometimes up to six) are born in April or early May. It is two weeks before they are well covered with fur, and a month before their eyes open. They are weaned at about five weeks, but may stay with their mother for up to a year. In the more southerly part of their range there may be another litter during the summer, but in the Northeast they are thought to have only one litter per year.

Habitat. Flying squirrels are found in heavily forested areas, including evergreen forests. The nest is usually built in an old woodpecker hole, and is constructed of lichens and shredded bark. Sometimes, however, these pretty little creatures will take up quarters in an attic.

Food. The diet is quite varied: acorns and other nuts, insects, seeds, berries, bark, fungi, tree flowers, and birds' eggs and their young.

Economic Importance. Except for the harmful insects it may eat, this is a rather innocuous little animal.

Comments. It is sometimes possible to see these little fellows in the daytime by scraping on the trunk of any tree that has a woodpecker hole or other cavity in it. If they are present, flying squirrels will usually poke out their heads to see what all the noise is about. (Needless to say, one has to scrape on scores of trees before being lucky enough to find flying squirrels by this method!)

Range in the Northeast. Throughout the northern sections but absent from southeastern Massachusetts, Rhode Island, southern Connecticut, southern New York and Long Island, and northern New Jersey

Similar Northeastern Species. The southern flying squirrel is smaller and not as brown on the back. It has pure white hairs on the belly.

Beaver *(Castor canadensis)* PLATE 17

Adult Size. Total length about 41″ (head and body about 28″, tail about 13″); weight usually 30–40 lbs., but sometimes to more than 60 lbs.

Description. This is the largest North American rodent, immediately identifiable by its broad, scaly, naked tail, which is horizontally flattened. It has webbing between the hind toes, and the second claw on each hind foot is split. The ears are short and rounded. The fur of both the upper parts and the underparts is a rich, dark brown.

Certainly the best-known aspect of the beaver is its skill as a builder. After a suitable site with a stream has been selected, the beaver proceeds to build a dam across the stream. This dam is constructed of logs, branches, rocks, and mud. As the water backs up behind the dam, the beaver extends the ends of the dam until a pond has formed. Some dams are extremely long—several I have found have been several hundred feet in length. Once the pond has formed, the house (lodge) is constructed out in the water. It is built up from the bottom, and is made of the same type of materials as the dam. It may protrude five or six feet above the water. Inside is a dry floor, usually with a warm bed of twigs, leaves, and grasses for the young. The entrances to this lodge are underwater, so that enemies cannot find their way in. Where a lake already exists, beavers may build their lodge on its bank, or even make do with a burrow within the bank.

Another of the beaver's engineering skills that is less well known is its canal building. This is a very slow-moving animal when on land, and although capable of defending itself with its enormous incisor teeth against small predators, it is easily killed by larger animals. To avoid exposing itself more than necessary it will, therefore, sometimes dig canals, radiating out from the pond, as a means of transporting branches from trees it has felled.

Beaver lodge

Trees are cut down with the large, orange-faced incisors. However, the beaver is far from being a skilled lumberjack, and is not able to make the tree fall in any direction it wants. There have been cases of beavers being killed by the tree on which they have been working.

The split claw on the second toe of each hind foot is used for combing the fur. The beaver usually does this soon after leaving the water, for it aids in waterproofing as well as in grooming. The large, flat tail is used as a rudder when swimming, with the webbed hind feet providing the propulsive power. The tail is also slapped on the water as a warning to other family members as the beaver dives away from approaching danger.

Since it does not hibernate, and since the water often freezes over in winter, the beaver has to lay in a supply of food for periods when it is unable to leave the pond. It accomplishes this by anchoring branches in the mud on the bottom of the pond.

The beaver's ears and nostrils have valves that close when the animal is underwater.

Breeding. Mating occurs during the winter, and two to six (usually three or four) young are born during April, May, or June after a gestation period of about 120 days. The young are born fully furred and with their eyes already open. They are able to leave the lodge and swim by the end of the first month. There is one litter per year, and when the kits reach about two years of age they either leave the colony (which is merely the one family group) voluntarily or are driven out by the parents. The adults mate for life.

Habitat. Beavers live in wooded areas where there are streams or lakes present. (In Alaska they are able to utilize alder in areas where there are no trees.)

Food. This animal lives on plant material. During the winter it eats the bark and tender twigs of poplar, willow, cherry, alder, birch, and (rarely) evergreen trees. In the summer it adds to this diet by feeding upon aquatic plants such as waterlilies and pondweeds, and it may also feed upon various grasses.

Economic Importance. During the 1700s and on into the nineteenth century, much of North America was first explored by beaver-trappers. The pelts were mostly exported to Europe, where beaverskin hats were very much in fashion. Millions of beavers were trapped for this thriving trade, and as a result their numbers declined tremendously; in fact, by the latter half of the nineteenth century the beaver had disappeared from much of the United States. Fortunately, strict protection other than during a carefully regulated trapping season has resulted in a great comeback in most areas. While beaver fur is still used for coats and trimming, this species is no longer in any danger of extinction. Ecologically, the beaver is probably of greatest value in its construction of ponds. Insects, fishes, frogs, salamanders, turtles, snakes, birds, and mammals take up residence in and around the ponds, thus creating large animal communities. In drought periods the ponds act as reservoirs for aquatic life, and in times of heavy rainfall both the ponds and their dams slow up the flow of water, and thus may prevent flooding and soil erosion farther downstream. On the debit side, a beaver dam may sometimes result in a stand of valuable trees being "drowned out," or it may even cause highway flooding. In such cases, the offending beavers are usually live-trapped by state conservation department personnel and taken to more remote areas where their labors will not bother anybody.

Although the flesh is quite palatable, it must be well cooked, for the beaver, like many other wild mammals, is subject to tularemia.

Range in the Northeast. Suitable habitats throughout all of the Northeast except Long Island

Similar Northeastern Species. The muskrat is smaller and has a slender tail that is flattened vertically. The river otter has a long, sinuous body and a well-furred tail.

Adult Size. Total length about 6.75″ (head and body about 3.75″, tail about 3″); weight about 0.75 ozs.

Description. The ears are large and naked, and the eyes are black. The hair-covered tail is shorter than the length of the head and body. During the summer the fur of the upper parts is light brown to rather orange-chestnut, especially on the top of the back, but in winter it becomes much grayer. The underparts are white throughout the year, the feet being white or light gray. The tail is brown above and white on the underside, but these colors are not sharply divided. The large eyes are a good indication that this is a nocturnal animal, rarely active during the day.

In addition to emitting sundry squeaks, these little creatures periodically utter a prolonged buzzy, humming sound, and may produce a faint drumming by tapping rapidly with their front feet.

Breeding. The breeding season is from early spring until the fall, during which time a female may give birth to several litters. The gestation period is about twenty-two days, and each litter consists of up to six blind, naked young. Their eyes open at two weeks, but it is another week before the young have a good coat of fur. They are able to mate at two months.

Habitat. While it is mostly a woodland mouse, this species is also to be found commonly along hedgerows and in areas where there is plenty of brush. In the late fall it often moves into houses to spend the winter months—my attic, for example, usually has a fair population throughout the cold weather. The nest is a snug, rounded structure made of almost any available material, including leaves, moss, shredded pieces of bark, grasses, hair, and chewed-up paper. The location is also varied: the abandoned nest of a squirrel, a hollow log, a bird nest-box, or inside a country house.

Food. Seeds and nuts seem to make up the bulk of the plant diet, and these are obtained from various grasses and sedges, dandelions, knotweed, oak, maple, pine, and wild cherry. Fruits such as blueberry and wild strawberry are also eaten in season. Animal food consists of insects such as beetles, moths, grasshoppers, and caterpillars, plus other invertebrates including centipedes, snails, and spiders. White-footed mice are also scavengers, and will eat from dead animals they may find. Supplies of seeds and nuts are hoarded against periods when food is hard to find.

Economic Importance. Although it helps in the control of weeds and some harmful insects, the greatest value to man is that, in acting as a "buffer species" (by being eaten by just about every predator),

Beaver

Porcupine

Muskrat

Eastern woodrat

Deer mouse

adult

young

White-footed
mouse

it keeps predators from possibly feeding on what are considered to be more valuable animals.

Comments. Several years ago my wife noticed that a good half of the fringe at one end of a blue scatter-rug in our bedroom had been neatly cut off. She wondered how this could have happened, but thought no more about it until more than a month later, when she pulled out a little-used drawer in the kitchen and there, tucked away at the back of the drawer, was a comfortable little white-footed mouse nest constructed entirely of the blue rug fringe. Since the kitchen is at the other end of the house from our bedroom, this must have been a very hard working mouse indeed! We conjured up a mental picture of this little creature running along with a mouthful of rug fringe, climbing up through the back of the cabinet to the drawer, incorporating this into the nest, and then going back for another load. While not really appreciated by us, this whole procedure would certainly have made a charming film sequence!

White-footed mouse's nest in a drawer

Range in the Northeast. Southern Maine and throughout the rest of the Northeast

Similar Northeastern Species. The deer mouse has a tail that is longer than that of the white-footed mouse (almost half the total length), with an obvious tuft of hairs at the tip. The tail is also much more sharply bicolored. The meadow mouse has smaller ears, a much shorter tail, very dark colored fur on the back, and a gray belly. The pine mouse has a very short tail and small ears. The boreal red-backed mouse has smaller ears, small eyes, and a shorter tail. There is a wide reddish stripe down the center of the back.

Adult Size. Total length about 5.5″ (head and body about 4″, tail about 1.5″); weight about 0.75 ozs.

Description. The ears of this stocky mouse are small and hairy. The eyes are small, and the tail is short. Fur on the upper parts is colored buff-gray, usually with a wide reddish stripe running along the top of the back. The underparts are pale gray. The coloring of the entire animal is brighter in winter than in summer.

Red-backed mice may be active during the day as well as at night. I once saw one in midmorning, busily feeding beneath one of my bird feeders. These mice often use the runways of other small rodents.

Breeding. From early spring through the summer and into early fall, the female may produce several litters. From two to eight young are born after a gestation period of about eighteen days. As with other mice, the young are born blind and naked, but grow so quickly that they are sexually mature at two or three months. The nest is usually lined with grasses, and is often in a burrow underground. It may also be built beneath rocks or logs.

Habitat. The red-backed mouse lives mostly in wooded areas, including evergreen forests, but it also inhabits sphagnum bogs and other wet habitats. In winter I have sometimes found these handsome mice in traps set in my garage, so apparently they are not averse to seeking shelter outside of their normal surroundings during cold weather.

Food. This species feeds to a great extent on roots, leaves, seeds, fungi, and berries. It stores large amounts of some of these materials in preparation for winter, and adds bark to its diet during periods of food scarcity. It eats few insects.

Economic Importance. While of some value in weed control, animals of this species sometimes nibble off the bark all the way around the trunks of young trees, "girdling" and therefore killing them.

Range in the Northeast. Throughout the Northeast except for Long Island

Similar Northeastern Species. The meadow mouse is grayer, lacks the reddish coloring along the back, and has a shorter tail. Pine mice have very short tails, and no red coloring, and both white-footed mice and deer mice have longer tails, larger ears and eyes, and white undersides.

Adult Size. Total length about 6.5" (head and body about 4.75", tail about 1.75"); weight up to about 2 ozs.

Description. Frequently referred to as a "mole," this is a rather stocky mouse with short, rounded ears, fairly small eyes, and a rather short tail. The fur is dense and quite soft. In winter it is a dark brown or gray on the upper parts, in summer, a lighter brown (but still very dark colored). A sprinkling of black hairs gives it a rather grizzled appearance. The underside is gray, sometimes a little buff.

Meadow mice are active by day and night throughout the year. Their activity usually occurs in short bursts of about two hours' duration, interspersed with resting and sleeping periods.

Although typically land animals, these mice swim well. One that I cornered on the bank of my pond at once turned and plunged into the water. It swam along for about twenty feet and then scrambled up into some tussocks and disappeared. They apparently do not climb as well as they swim. I used to allow my tomatoes to ripen on the ground, and lost much of the ripe fruit to meadow mice living beneath the hay mulch. Once I began staking up the plants this was no longer a problem.

Breeding. Meadow mice are among the most prolific of mammals. During the breeding season, from spring well into November (and occasionally right through the winter) there is a succession of litters. (A captive female is reported as having had seventeen litters in a single year!) The gestation period is twenty-one days, and each litter is made up of three to nine blind and naked young. The baby mice are weaned before they are two weeks old, and females are able to breed at the age of three weeks. Males breed a week later.

Habitat. This species prefers areas either where it is moist or where there is a lush growth of grasses. Thus, it is most common in wet meadows, hayfields, and other grasslands, even along the coast. However, it may be found almost anywhere in the countryside. The nest is usually located on the ground, in the midst of a tussock of grass, under a rock, or underground. I usually find a few beneath the hay mulch I put between my vegetable rows. Nests are compact balls of dried grasses and roots, sometimes lined with mosses or other soft material.

Food. Meadow mice eat a tremendous variety of grasses and sedges and their seeds, roots and tubers, bulbs, flowers, leaves, and bark. They also eat some insects and carrion. In their search for food these little creatures habitually follow the same routes. They may nibble off the grasses along the way, and these little trails become easy to spot in the fields and along the hedgerows. Each path is about 1.5" wide and has branches radiating out through the grass.

Economic Importance. Because of their great numbers, wide variety of habitats, and extremely varied diets, meadow mice are probably the greatest rodent crop pests in the Northeast. In the farmers' fields they eat great quantities of grains and other crops, and in gardens they ruin many vegetables. In my own garden they have feasted upon potatoes, carrots, turnips, muskmelons, and tomatoes, so that in some years a good quarter of these crops has been useless. While I do not object to giving them bed beneath the mulch, I certainly resent giving them board also, and have accordingly carried out extensive trapping campaigns against them.

After one particularly harsh winter I noticed that in a row of five-foot white pine trees, half a dozen had suddenly gone brown and died. Upon examination I discovered that each one had been girdled at ground level. A few traps set nearby confirmed my suspicion that meadow mice were the culprits. Similar girdling in a planting of young fruit trees would have serious effects in an orchard.

These mice are food for almost every type of predator, including foxes, raccoons, weasels, skunks, bobcats, hawks, owls, and even crows. Thus, they are important links in many food chains.

Range in the Northeast. Throughout the Northeast

Similar Northeastern Species. The pine mouse has a tail no longer than one inch. The white-footed mouse and the deer mouse both have large eyes and ears, white undersides, and tails that are about three inches long. The boreal red-backed mouse has a broad reddish stripe down the center of the back, and is generally browner. The yellow-nosed mouse has a distinctly yellow nose.

 Muskrat *(Ondatra zibethicus)*   PLATE 17

Adult Size. Total length about 22" (head and body about 12", tail about 10"); weight 1.5 to 3 lbs.

Description. This big, stocky, short-eared rodent has a tail that is practically naked and that is flattened vertically. The legs are short but the hind feet are quite large, with the toes partly webbed. The dense underfur is gray, but is covered with long, rich brown "guard hairs" so that the overall color of the animal is brown except for a whitish throat and belly. The tail is very dark.

Muskrats are active all year long. Although feeding mostly in the evening and at night, they may often be seen during the day also. On Long Island I have many times seen them swimming in coastal marshes under the midday sun. They swim by "sculling" with their vertically flattened tails and paddling with their hind feet.

Although I have heard a low-pitched, moaning sound from a muskrat, this has been only when the animal was caught in a live-trap. At such times it also clicks and grinds its teeth.

Breeding. In the Northeast, two or three litters are produced from the end of April through the summer. The gestation period varies, but is usually about thirty days. Each litter averages five or six young. Although blind at birth, the four-inch-long babies have a thin covering of hair. This hair grows rapidly, and by the time their eyes are open at two weeks it is quite thick and grayish in color. The young are weaned when a month old, and then fend for themselves. At this age their fur looks quite black when they are seen swimming. Should they not leave home voluntarily, they are then driven away by their mother, who is now preparing to give birth to her next litter.

Muskrat house

Habitat. Although primarily found in shallow fresh water and salt marshes, the muskrat also occurs along the banks of streams and ponds, and even in swamps if there is plenty of food. It builds a conical or domed house of various water plants. This is constructed in water from one to two feet deep and is based upon subsurface material scraped together to form a pile on the bottom. The house usually protrudes about two feet from the water. Inside is a warm room just above the water level, with one or two entrances. Smaller "feeding platforms," which provide shelter from enemies, may also be built here and there. Along streams, or in deeper ponds or lakes, the muskrat becomes a bank burrower. It digs a tunnel leading from about a foot below the water level up to within a few inches of the surface of the ground. The tunnel measures about six inches in diameter and leads to a sleeping chamber. Usually another escape tunnel is dug out from this nest.

Food. Aquatic plants are the main food items: cattails, water lilies, arrowheads, various pondweeds, and sedges. To a lesser extent crayfish, freshwater clams, frogs, and dead or dying fishes are eaten.

Economic Importance. As a species, the muskrat is the most important fur-bearer in the United States; well over 5 million skins are marketed each year. These days the fur is used mostly for trimming coats, but muskrat coats are also still made.

When it burrows into banks the muskrat may do much damage, especially if the "bank" happens to be an earthen dam!

Muskrat meat is very tasty, and when roasted and served cold it reminds me a little of chicken.

Comments. One does not actually have to see a muskrat—or even a muskrat house—to know that this species is present in an area. At my pond, where muskrats periodically take up residence, there are several indications of when they are there: their oval droppings appear in small heaps on boulders along the shoreline, cattail leaves will be chewed to the water level, and pieces of cattail stems and leaves can be seen floating along the edges of the pond.

If one has a pond with an earthen dam (as I do) there is a real risk of muskrats weakening the structure with their tunneling.

Predators such as mink, harriers, and great horned owls are constant threats to muskrats. The former is especially feared, for it will pursue a muskrat right into its home.

Range in the Northeast. Throughout the Northeast wherever living conditions are suitable

Similar Northeastern Species. The beaver is much larger and has a large, paddle-shaped tail flattened _horizontally_. The mink has a long, sinuous body and a fairly bushy tail, and the otter is larger, with a wide snout and a thick, furry tail.

Adult Size. Total length about 15″ (head and body about 8″, tail about 7″); weight about 12 ozs., and sometimes half again as large

Description. The outstanding features of this unpopular animal are large, naked ears and a long, scaly, almost hairless tail. The fur is coarse-looking, brown to gray on the back and light gray on the underside. The tail is gray, and the feet buff or gray.

Although often seen during the day, this is mostly a nocturnal mammal. It is an adept swimmer and climber, and is quite vociferous. Especially when it is angry, it produces a variety of squeaks and squeals.

Breeding. This is a very prolific creature. It breeds throughout the year, although most litters are produced during the warmer months. Each of the four or more litters per year consists of approximately seven young, and the young can breed at the age of three months. (Thus, theoretically, a single pair of Norway rats and their offspring would produce several million rats in the space of three or four years were it not for natural controls such as disease and predators.) The gestation period is about twenty-two days. The newborn blind and naked young open their eyes after two weeks and are weaned at three weeks.

Habitat. Norway rats are found in most areas where there is plenty of food, water, and cover. Colonies are common in all large urban centers, in farming country, and in areas where livestock are housed. The rats tunnel beneath or close to buildings. Many rats may use the same runways, and even the same nests. These nests are built along the runways, and are usually about twelve inches in diameter and lined with any soft material close at hand: paper, cloth, grasses, etc.

Food. This voracious creature eats almost anything: all manner of grains and vegetables, meats, fruits, soap, eggs (and often chicks and chickens too!), leather, paper, cloth, garbage of all kinds, and carrion.

Economic Importance. It is often said that the Norway rat has been responsible for more human deaths than all of the wars in human history added together. This is probably true, for a diseased member of this species, or the fleas and lice that live upon it, can transmit to man several deadly diseases: bubonic plague, typhus, tularemia, trichinosis, and jaundice, to name but a few. When one considers that a single epidemic of bubonic plague (or "the black death," as it was formerly called) once wiped out about a quarter of the total

Meadow mouse

Pine mouse

Yellow-nosed mouse

Boreal
red-backed mouse

Bog lemming

Woodland jumping
mouse

Meadow jumping
mouse

House mouse

Norway rat

human population of Europe, and that such epidemics were once fairly frequent in other parts of the world, the deadly potential of this animal can be well realized.

As well as carrying disease, the Norway rat does much damage to foodstuffs, and eats or ruins both standing and stored crops. It gnaws on buildings and furnishings, and sometimes causes fires by nibbling off the insulation from electric wiring.

Man makes use of these animals by using them experimentally in hospitals and laboratories (white rats are albino Norway rats), and many predators utilize them as food. Other than this there is little that can be said in the rat's favor. If one were to exclude man it is potentially the most dangerous mammal in the world.

Comments. This is an accidental import into the United States. It is believed to have originated in Asia, and to have moved from there into Europe by hiding aboard ships. From Europe, Norway rats found their way to America by the same method, and are thought to have reached here during the 1770s. Since then they have become established throughout the country.

In addition to its most common American name, "Norway" rat, this animal is known by many other names—for example, brown rat, house rat, wharf rat, river rat, sewer rat, gray rat, barn rat, and roof rat.

Range in the Northeast. Throughout the Northeast, particularly in areas heavily occupied by man

Similar Northeastern Species. The eastern woodrat has a furry tail and white underparts. The muskrat is much larger and stockier and has a tail that is flattened vertically.

House Mouse *(Mus musculus)*   PLATE 18

Adult Size. Total length about 6.5" (head and body about 3.25", tail about 3.25")

Description. This small mouse has large ears and a long, almost hairless tail. It is colored gray-brown above, with a paler, buff underside. It climbs and swims well, although preferring to avoid water whenever possible. It is active mostly at night. The house mouse can make a series of squeaks, and actual songs, described as a weak trilling that can be heard only at close range, have been often reported.

Breeding. Like the Norway rat, the house mouse has a very high breeding potential. About six litters are produced each year, and each litter averages five or six blind and naked young. They remain with their mother for about three weeks, by which time they are weaned. Three weeks later they are able to breed! The gestation period is about twenty days.

Habitat. Although usually found in buildings within urban areas, these mice sometimes turn up in fields and croplands in large numbers. In an urban area the warm, cosy nest is often built of scraps of cloth and chewed-up paper, but those living closer to the countryside may use grasses.

Food. This little animal will eat almost anything. It prefers grains and other seeds but will feed upon any of man's foods (meats as well), or on soap, candles, leather goods, and the bindings of books.

Economic Importance. While it does not normally transmit disease to man, this species is sometimes known to carry human disease organisms. In addition to the food they eat, house mice render much more food unfit for human consumption, and their droppings and urine leave a strong, unpleasant odor. Like Norway rats, they do much damage to buildings and furnishings with their gnawing. The white mice that are used by the millions in experimental work and for feeding captive animals are actually albino house mice.

Comments. This is another Old World animal that reached America by stowing away on ships. It is thought to have reached here at about the same time as its cousin the Norway rat—roughly in the 1770s.

Range in the Northeast. Throughout the Northeast, especially in large cities

Similar Northeastern Species. All of the other Northeastern mice with large ears and long tails like the house mouse differ from it in having white bellies.

Meadow Jumping Mouse *(Zapus hudsonius)* PLATE 18

Adult Size. Total length about 8.25″ (head and body about 3.25″, tail about 5″)

Description. The very long, tapering tail (much longer than the head and body) and the long hind legs at once identify the jumping mice. The ears, while fairly small, are quite easily seen. The fur of the meadow jumping mouse is long and coarse, and is colored yellow-brown along the back, with a darker band running down the top of the head and body. The sides are more yellow, the underparts and feet white. The tail is dark along the top and white on the underside, with little hair. There is a small gray tuft at the tip of the tail.

While the normal gait is a series of small hops, the large hind legs enable this little creature to make leaps of several feet. At such times, the tremendously long tail acts as a balancing organ—in much the same way as the tail of a kangaroo. Jumping mice climb well and are good swimmers (though only the hind legs are used in swimming). The voice is said to be a rather low-pitched note, although I have never been lucky enough to hear it.

Along with the woodchuck and some of the bats, the jumping mouse is a true hibernator. By the end of September it has usually moved into its winter quarters, a snug, grass and leaf nest dug in the ground well below the frostline. Here it curls up tightly and sleeps, seldom emerging until early May. During this long period of sleep it is sustained by the heavy layer of fat that it built up in the late summer, which its body consumes very slowly. This is accomplished by a marked slowing of its metabolism; the heartbeat and breathing slow down, and the body temperature drops radically.

Jumping mouse in hibernation

Breeding. There are usually two litters per season, one in June and the other in late summer. Litters usually consist of about four young, born blind and naked, and having short tails. By the end of three weeks their eyes are open, they are well covered with fur, and their tails have grown to almost full length relative to their body size. At six weeks they are fully grown. The gestation period is somewhere between three and four weeks in length.

Habitat. These jumping mice are usually found in fairly moist grassland, where the grass is long, but I have several times seen them bounding away into the hedge when I have been mowing my lawn.

Food. As might be expected from their favored habitat, these pretty little creatures feed on a wide assortment of seeds, grasses, and berries, and also some insects. I once saw one balanced precariously in a high-bush blueberry, reaching carefully for the fruit. Unfortunately it spotted me and leaped down into the grass before I was able to see just how many berries it could consume. Most rodents store food for the winter months, but the fact that these animals hibernate makes this hoarding unnecessary.

Economic Importance. Probably this little animal is of some value in insect control, but on the whole it is an innocuous creature.

Range in the Northeast. Throughout the Northeast in suitable habitats, but in "pockets"

Similar Northeastern Species. The woodland jumping mouse looks very much like this species, but is more brightly colored and has a distinct *white* tuft at the tip of the tail. None of the other Northeastern mice have such tremendously long tails or such large hind legs.

Adult Size. Total length about 34″ (head and body about 28″, tail about 6″); weight 10 to 20 lbs., occasionally heavier

Description. The porcupine is a stocky, heavy rodent with a small head, short ears, legs, and tail, and a very hunched posture. The general color is black or very dark brown with a frosting of white- or yellow-tipped hairs. Scattered throughout the fur are many whitish quills with dark tips. These are very sharp and have tiny barbs at their tips. They are particularly numerous on the tail and at the rear of the body. When attacked, a porcupine will turn its back, raise its quills, and thrash its tail vigorously. Contact with the quills causes them to become embedded in the flesh of the attacker, and the barbs at their tips help pull them free from the porcupine. They then work deeper into the flesh of the enemy, causing intense pain and sometimes death.

This animal moves at slow speed, but can run clumsily for short distances. Its vision is poor. I have moved quietly up to within a few feet of feeding porcupines before they have become aware of me.

Porcupines are powerful swimmers. I once saw a large porcupine wade into a broad, very fast flowing stream and swim across. The current swept it a small distance downstream, but it doggedly battled its way toward me and eventually climbed out not too far from where I was standing. Although they are good climbers, these mammals move about quite slowly in trees, even if alarmed.

Although they are usually silent, I have heard a variety of sounds from them. When alarmed, they will sometimes click their teeth and utter a low, groaning sound. One night on a camping trip I heard an eerie moaning near my tent, and on investigating saw two full-grown porcupines. One of them was lumbering after the other one, uttering the moaning I had heard. Presumably this was some sort of courtship behavior, for the one being followed did not seem alarmed.

Breeding. After a gestation period of about 210 days a single young is born somewhere between March and early May. Relative to the size of the mother it is a huge baby, weighing about one pound. It is born with fur and with its eyes open. Although it arrives with its quills already developed, they do not harden until dry—otherwise the mother would be in real trouble! Weaning is a rapid process, and by the time it is two weeks old the young porcupine is eating plant food. It remains with its mother for about six months.

Habitat. Although it is sometimes to be found living in brushy areas, the preferred habitat certainly is well-wooded regions. The den is often in a hollow tree, or maybe in a hollow log on the ground. Sometimes a cave or rock crevice is selected.

Food. During the warmer part of the year porcupines feast on a wide range of plant material: leaves, berries, grasses, and even aquatic plants. In winter they feed on the bark of hemlock, spruces, pines, and many deciduous trees. They are very fond of salt and will gnaw on any sweat-soaked leather or wood. They also gnaw on antlers shed by deer or moose, although this is probably to fill a need for calcium.

"Girdling" on a tree

Economic Importance. Because of its liking for the inner bark of trees, this animal sometimes does much damage by "girdling." If the girdling should take place on the trunk, the entire tree will die above the point where the bark has been nibbled away. In valuable stands of lumber trees or in orchards, this sometimes becomes a serious problem.

Anything with a salty flavor may be gnawed by a porcupine. Horse harnesses and buggy shafts in porcupine country are particularly vulnerable if they are not left out of reach.

The quills were once widely used by Indians to decorate clothing and various other articles. They are sometimes still used in the same way today.

Comments. Although often called "hedgehogs," porcupines are completely different from those strictly Old World animals. True hedgehogs are members of the order Insectivora (which includes shrews and moles), rather than being rodents, and have unbarbed spines instead of quills.

Folklore. The porcupine is commonly supposed to "fire" its quills at an enemy. This is not true, but because some of the older quills are loosely attached they may accidentally fly off when the animal thrashes its tail around.

Range in the Northeast. Throughout the more northern sections, but absent from eastern Massachusetts, Rhode Island, most of Connecticut, extreme southern New York (including Long Island) and New Jersey

Similar Northeastern Species. None. Its clumsy movements and its numerous quills at once separate this large rodent from all other Northeastern mammals.

Other Northeastern Rodents

Fox Squirrel (*Sciurus niger*)   PLATE 16

The fox squirrel is about 21″ long (head and body about 11″, tail about 10″) and weighs about two pounds. Somewhat like the gray squirrel in appearance, this larger animal has brownish gray coloring on its back, orange-brown ears, and pale, orange-brown underparts. The tail has a mixture of black and brown hairs. In the Northeast it is found only in extreme western New York.

Southern Flying Squirrel (*Glaucomys volans*)   PLATE 16

The total length is about 9″ (head and body 5″, tail 4″), the weight about two ounces. Although like the northern flying squirrel in most respects, this smaller animal has completely white belly hairs. (The northern flying squirrel has belly hairs that are lead colored at the base.) Its habits are much the same as those of the northern flying squirrel, and it is found throughout the Northeast except for northern New Hampshire and Vermont and all but extreme southern Maine.

Deer Mouse *(Peromyscus maniculatus)*   PLATE 17

Total length about 6.75″ (head and body about 3.5″, tail about 3.25″). Deer mice resemble white-footed mice but have tails that are nearly the same length as the head and body and that are much more sharply bicolored (dark brown above, white below). They are missing from eastern Massachusetts, Rhode Island, eastern Connecticut, extreme southern New York (including Long Island), and northeastern New Jersey.

Eastern Woodrat *(Neotoma floridana)*   PLATE 17

This animal is about 16.5″ long (head and body about 9″, tail about 7.5″) and weighs about 14 ounces. Although rather like a stocky Norway rat at first glance, the furry, bicolored tail at once serves to identify this native rodent. This is the so-called "pack rat," named for its habit of accumulating piles of trash in its nest. The nest itself is a fairly large structure, usually made of twigs and built in a crevice or on a rock ledge. It may contain pebbles, bones, shiny metallic objects, and all manner of other rubbish collected by the rat. The food consists mostly of fruits, nuts, berries, fungi, and other plant material. In the Northeast, woodrats are found only in parts of southern New York (but not Long Island) and western Connecticut.

Bog Lemming *(Synaptomys cooperi)*   PLATE 18

The total length of the bog lemming is about 5″ (head and body about 4.25″, tail about 0.75″). This is a small, plump, mouselike rodent that is brown-gray above and grayish on the underside. It has a very short tail and short ears. The fur is fairly long and coarse. Bog lemmings live in low-lying wet areas where there is long grass. Although it is not common in the Northeast, this species is found throughout the area except for eastern Massachusetts, Rhode Island, southern Connecticut, extreme southern New York (and Long Island), and northeastern New Jersey.

Yellow-nosed Mouse *(Microtus chrotorrhinus)*   PLATE 18

This mouse is about 5.75″ long (head and body 4″, tail 1.75″). In appearance and coloring it is rather like the meadow mouse, but it has a very yellow or orange nose. This is an uncommon species found at higher elevations in shaded, moist, rocky woods. Its range in the Northeast is a belt running from northwestern Maine down through northern New Hampshire and Vermont, and northeastern New York south through the Catskills and into northeastern Pennsylvania.

Pine Mouse *(Microtus pinetorum)*   PLATE 18

The pine mouse is about 3.5″ long (head and body about 2.5″, tail about 1″). It is a small, chunky, short-tailed animal with very small ears. The back and sides are a shiny, dark rust in color; the belly is a light gray. The tail is brown, paler on the underside. This little mouse is found in almost any type of habitat, but it prefers wooded sections where there is plenty of loose soil in which to burrow. Pine mice are great pests in orchards, for they nibble the bark from the roots of the fruit trees. In the Northeast they are not present in Maine, northern New Hampshire and Vermont, and much of northern New York.

Woodland Jumping Mouse *(Napaeozapus insignis)*   PLATE 18

This creature is very much like the meadow jumping mouse in size and color pattern, but it is much more brightly yellow and has a distinct white tuft at the tip of its tail. As is suggested by its name, its favored habitat is in forest country, and it is rarely found far from wooded areas. Its habits are much the same as those of the meadow jumping mouse. It is absent from eastern Massachusetts, Rhode Island, southern Connecticut, extreme southern New York (including Long Island), and northeastern New Jersey.

Rabbits and Hares

The Timid One

Night is ending. I see a line of gray creeping into the sky at the eastern horizon, and the breeze dies away. The air is cool and fresh and clean. From somewhere along the black hedge-row a robin bursts into song, the cheerful notes echoing out through the darkness.

The light grows slowly. From being mere shapeless blocks of shadow, objects begin to take form. They become trees and shrubs, rocks and clumps of ferns. Other birds add their songs to that of the robin: song sparrows, redwing blackbirds, field sparrows, and meadowlarks. A bobolink joins the growing chorus. Here is one of the most exuberant of all bird songs, a loud, rollicking succession of notes that seems to dance through the air, filling it with warmth and vitality—a song that brings a smile to the morning.

Now it becomes still lighter, and all at once I realize that what I have taken to be a small rock at the edge of the field is actually a rabbit. The rabbit is sitting hunched up in the wet grass. Through my binoculars I see that it is munching dandelion leaves, its nose twitching rapidly as it chews. As I watch, it stretches its neck and unhurriedly hops forward to another clump, selecting a leaf, nibbling daintily at the tip and then working quickly down the length of it. It begins on another, but hears something moving in the hedge nearby. Instantly the rabbit freezes. Its chewing stops and it crouches low in the grass. Whatever has alarmed it—perhaps a foraging chipmunk—goes its way, and presently the rabbit resumes its feeding. But it is plainly nervous now, and frequently stops eating in order to survey the hedge. Apparently its fear eventually gets the better of it, for with a series of cautious hops it begins to move further out into the field and away from the hedge. It comes directly toward me.

As it approaches I continue to watch it through my binoculars. Suddenly it sees me. It leaps into the air, twisting its body as it does so, and streaks back toward the hedge. As it bounds over the grass, veering from side to side, its white tail flashes. It reaches the hedge and is gone.

Rabbits are charming creatures—warm and cuddly-looking, beloved by children and smiled upon by adults. But, as every gardener knows, there is another side to them. One who has seen an entire row of lettuce seedlings demolished overnight by a rabbit will be inclined to smile no more upon that rabbit—or any other rabbit for that matter. One who, like myself, has lost young ornamental trees in winter, due to girdling by hungry rabbits, may take a rather dim view of their entire tribe. They are, indeed, among the most destructive of garden pests, and in the past have been the source of much aggravation to me, the objects of a great deal of execration.

Yet, when all is said and done, rabbits cannot be blamed for needing to eat. If I was foolish enough to make available to them tasty salads and other delicacies, I have only myself to blame if they take advantage of this bounty. My garden is now protected by a fence, and my trees by wire mesh cages. I no longer have a problem. Rabbits and I exist in harmony, and I am able to enjoy them as neighbors, rather than cursing them as thieves and vandals!

General Information

Although they were once considered to be a specialized group of rodents, rabbits and hares are now placed into a separate order—Lagomorpha. One of the ways in which they differ from rodents is that they have two pairs of upper incisor teeth (gnawing teeth), as well as a pair in the lower jaw. One of the upper pairs is located as are those of rodents; the second pair occurs as small, peglike teeth directly behind these. Rabbits give birth to blind, almost naked young, while hares are born with their eyes open and a good coat of fur. Members of this group have always been important to man for their meat and skins. The hair is also used for making felt.

Adult Size. Total length about 18.5″ (head and body about 17″, tail about 1.5″); weight about 3 lbs.

Description. This medium-sized hare (often wrongly called snowshoe *rabbit*) has very large hind legs and feet, long ears, a short tail, and very dense fur. This is one of the mammals that changes its color with the seasons. In summer the fur is mostly gray-brown, with darker hairs along the back and on the hindquarters and with some chestnut on the sides of the head. The chin, throat, belly, and underside of the tail are white; the upper side of the tail is black. In winter the entire animal becomes white except for dark tips to the ears. This seasonal color change gives it its other name of "varying hare."

The snowshoe hare is nocturnal. During the day it stays within heavy brush or sits quite still in a shallow depression it has hollowed out with its body. (This depression is known as a "form.") If alarmed it can move very rapidly. Even in deep snow its large, furry feet spread out its weight so that it can move at great speed without becoming bogged down. It swims well, but will not enter water without good reason.

It will sometimes utter grunting sounds, and will thump on the ground with its hind feet. Like all members of its group, it screams shrilly when injured or frightened.

Breeding. From one to six young (usually three or four) are born from late March through early August. The gestation period is thirty-six days, and there may be two or three litters during the season. The young of hares are known as leverets. They are born well-furred, and with their eyes open. There is no nest. The leverets are nursed for about four weeks and are then able to feed themselves on plant food.

Habitat. Snowshoe hares live in places where there are dense forests or swamps, but sometimes they may be found on brushy hill slopes. The home range is only about five acres in extent, but this animal may temporarily travel away from it.

Food. The main food during the summer consists of grasses, dandelions, buds, berries, and other juicy vegetable matter. In winter, when these items are mostly unavailable, the snowshoe hare switches to a diet of tender twigs, conifer needles, and bark. Pines, spruces, white birch, and aspen are favorite trees. These hares are also known sometimes to gnaw upon dead animals they may find.

Economic Importance. By girdling trees in winter, this hare may kill many seedling trees and saplings. It is a valuable game animal, and

New England cottontail

Eastern cottontail

Snowshoe (varying) hare

winter

summer

European hare

is used as food in many areas. The pelts are used in the fur trade, although they are not very durable.

This is an important prey species for lynx, bobcats, foxes, and weasels.

Comments. Like many animals, the snowshoe hare has population cycles. In its case these cycles take approximately ten years. A population peak is reached, after which the hares decrease until few are to be found. Then their numbers gradually build up again to the next peak.

While camping out in the Canadian woods in early June, I was startled to hear a tremendous racket among the dead leaves one night. When I emerged from my tent to drive away what I was sure was a bear, I saw a snowshoe hare leaping and tumbling in the moonlight. Whether this was some form of courtship display or merely high spirits, I do not know, but I was greatly relieved to discover such a harmless intruder!

Range in the Northeast. Although largely trapped out from many areas, the snowshoe hare has been reintroduced by various state conservation agencies; now becoming fairly common in many of its original haunts, and being found in most of the Northeast except for eastern Massachusetts, parts of Connecticut, extreme southern New York (including Long Island) and northeastern New Jersey

Similar Northeastern Species. In its summer coat, this species may be taken for a cottontail. Cottontails, however, have whitish feet, not the brown of the snowshoe hare. The European hare is very much larger, has longer ears, and is found in open country rather than in woods.

Eastern Cottontail *(Sylvilagus floridanus)*   PLATE 19

Adult Size. Total length about 16″ (head and body about 14″, tail about 2″); weight 2–3 lbs.

Description. The ears of this fairly large rabbit are about 2.5″ long. The soft fur is brown to gray on the upper parts, with a sprinkling of black-tipped hairs. The underparts are white. There is a reddish patch at the nape of the neck, and the rump is appreciably grayer than the rest of the upper parts. The feet are off-white, the tail brown above and white below.

While active in late afternoon and early morning, this is mostly a nocturnal animal. It is very timid, and is a rapid runner.

Breeding. The cottontail is a prolific creature, giving birth to several litters of three to six young each year. The first of these litters is usually produced in March, the last in early September. The gestation period is about twenty-eight days. The female digs a shallow depression (or sometimes uses a natural hollow) and builds a warm nest of chewed up grasses and leaves lined with soft hair from her own underparts. When leaving this nest, she invariably arranges leaves or long grass over it so that it is well concealed. The young are blind and almost hairless at birth, but their eyes open in a week. After little more than two weeks they leave the nest for good. Cottontails normally breed at about six months of age.

Cottontail rabbit's nest and young

Habitat. Quite diverse areas may be selected in which to live: heavy brush, the edges of fields or woodland, wet wooded areas, suburban gardens and even city parks if there is enough cover present. The home range covers several acres, but this may be shared with other rabbits.

Food. During the warmer seasons the cottontail finds plenty of food in the form of many grasses and broad-leafed green plants such as dandelions, plantain, sheep sorrel, and wild strawberry. In winter these plants are unavailable, and the diet is then made up of the twigs and bark of a large variety of trees and shrubs, including sumac, elderberry, gray dogwood, wild grape, hawthorn, buckthorn, various fruit trees, and sapling maples and birches.

Economic Importance. This is one of the most important game animals in the Northeast. (Rabbits killed for food should be well cooked. They harbor many internal parasites, and are also known sometimes to transmit anthrax and tularemia.)

Although attractive looking, and therefore often used in coats and trimmings, rabbit fur is one of the least durable of furs.

Rabbits sometimes feed in vegetable or flower gardens, and in winter they may kill many young trees by girdling. They themselves are fed upon by many animals: lynx, bobcat, foxes, coyotes, weasels, and the larger species of hawks and owls. The young are eaten by almost any meat-eater—including large snakes—that may discover their nest.

Comments. Like the snowshoe hare, cottontails have peak population periods, although they seem to be less well defined. On my farm they seem to reach a peak every six years or so, but this may not be accurate for the entire Northeast.

Snowshoe hare tracks in the snow

Range in the Northeast. New Jersey, Pennsylvania, Connecticut, most of New York south of the Adirondacks (including Long Island), most of Massachusetts, Rhode Island (see also New England cottontail).

Similar Northeastern Species. The New England cottontail has a black patch between the ears and is much redder in color. The snowshoe hare in its summer coat lacks the reddish nape patch of the eastern cottontail and has brownish feet, while the European hare is very much larger and has longer ears.

Other Northeastern Rabbits and Hares

New England Cottontail *(Sylvilagus transitionalis)*   PLATE 19

This rabbit is approximately the same size as the eastern cottontail, but is a little redder in coloring and has a black patch between the ears. Its habits are essentially the same as those of the eastern cottontail. It ranges from southern Maine through all of the New England states, eastern and southern New York (including Long Island), and into eastern Pennsylvania and northern New Jersey.

European Hare *(Lepus europaeus)*   PLATE 19

The European hare has a total length of more than 24″ and weighs up to ten pounds or more. This very large introduced hare is light brownish gray in coloring. It has a black-topped tail and black tips to the ears. It can travel at speeds of more than forty miles per hour.

Almost as soon as it was introduced into southern New York, early in the twentieth century, people realized that its value as a game animal was far outweighed by the damage it can do by girdling orchard trees. Efforts are still being made to exterminate it, but it can still be found in the lower Hudson River Valley. This is an animal about which there are many legends. The fable of the "Easter bunny" is thought to have come down to us from several thousand years ago. In Europe, at that time, it was said that hares were once birds, and that they were changed into their present form by the goddess of spring. As a mark of thanks, these animals laid eggs at the festival of the goddess, in the spring.

When I worked on a farm in England, combine harvesters were a rarity. It was the usual practice to make stacks of harvested wheat, and then to thatch them as protection from the weather until they could be threshed. The thatcher would often construct a straw hare at each end of the stack's peak. This was regarded as a lucky symbol. While he undoubtedly did not realize the true significance of this, it is a fact that, long ago, the hare was believed to be the spirit of wheat.

The "Mad March Hare" of *Alice in Wonderland* is a character that got its name from the strange behavior of hares during their courtship season in March. At this time the males leap and run around in extremely erratic fashion.

Deer

A Winter Tragedy

Late on a January morning I leave the house and head for the pine woods on top of the hill. The ground has been frozen hard for several weeks, and ice is thick upon the pond. Several nights ago some eight inches of snow fell and whitened the countryside. A slight thaw followed, and the upper layer of snow began to melt. Then the temperature dropped again. The wet surface snow froze, and a crust formed. Yesterday there was more snow, which covered the hard glaze so that it lies two or three inches under the new snow.

Now a savage storm is lashing the landscape. A biting wind howls across the hillside and swoops out over the open fields. The wind-blown snow streams from the ground like spume from a wave crest and whirls and hisses away in blinding crystal sheets. As I push my way through a belt of bare, brittle sumacs, I plunge to the knees in snow that has drifted in under the branches. There are still some fuzzy red cones of fruits at the tips of some of the branches, and these will serve as food for grouse and other birds. Near the surface of the snow the bark of many of the sumacs has been nibbled off by hungry rabbits.

Behind the sumacs are the pine woods reaching back over the crest of the hill. When I gain their shelter the sounds of the storm are dulled, but the crowns of the trees sway wildly, and boughs rub and groan against each other as the wind whistles and whines through the upper branches. The wind brushes the snow from the green needles, and the cold air is filled with drifting white granules. The lowest tree limbs are anchored to the ground by the weight of snow upon their foliage. They form small caves that are used as snug refuges by many small creatures. Before the storm the tracks of mice, rabbits, grouse, and squirrels patterned the snow near these shelters, but now the wind has obliterated most of them.

In one corner of the woods, where the trees grow very closely together, is a fallen white pine. Its roots, and the soil frozen between the roots, form a solid, rounded wall four or five feet high. In the hollow where the tree used to stand I find a deer. The deer has obviously been slowly starving for many days. The outline of its rib cage is plainly visible beneath the snow-encrusted, gray winter fur. The sides of its body are deeply hollowed. Its breathing is labored. Sometimes its bony legs twitch spasmodically and its eyes roll whitely, but otherwise it lies still.

In my mind's eye I trace the story of the deer's downfall. When winter came it could no longer find the leaves, weeds, and grasses on which it had been feeding. It began eating tender seedling trees and the twigs of maples, poplars, willows, and sumacs. For a time this was sufficient. But there are many other deer in our area, and steadily this food source shrank. The seedling trees were consumed and the twigs, too, for as high as the deer was able to reach. In desperation it gnawed bark from the large trees and munched the dry canes of wild raspberries. This put something into its empty belly and helped to reduce the hunger pains, but there was little nourishment in these materials. They did nothing toward replacing the fat that was now almost gone. Then the snow came, and movement became difficult. As it finally limped into the shelter of these woods its legs rubbed against the sharp edges of the crust, and I can see raw, bloody patches on them.

But while there is shelter here, there is no food. After a time, as its body fires burned lower and lower, the deer found this hollow behind the roots of the fallen tree. Too weak to search further for food, it probably stood here for a while and then sank to the snow. Eventually it stretched out on its side, as I now see it.

Now it is near death. The snow drifts and blows and eddies around it, and the wind moans through the creaking branches above its head.

General Information

Members of the order Artiodactyla are even-toed hoofed animals (that is, in the case of the deer family, "cloven hoofed"). Our Northeastern representatives are large animals, with the males having antlers. (These should not be confused with horns. Antlers are composed largely of calcium—like bone— and are shed each year. Horns are of keratin and remain on the animal throughout its life. The farmyard cow is an example of a horned animal.) Members of the deer family are mostly browsers, feeding on leaves and twigs, but may also do a fair amount of grazing. They are cud-chewers. The food is swallowed rapidly and then later on, when the animal is lying quietly, regurgitated into the mouth in small masses. This time it is properly chewed before being reswallowed for complete digestion. These animals have always been important as game species. In the early days of settlement their skins were used for making clothing, and their meat was eaten. Today they are hunted only for their meat and for the "sport" they provide. The only members of this order in the Northeast are the white-tailed deer and the moose.

White-tailed Deer *(Odocoileus virginianus)* PLATE 20

Whitetail deer tracks

Adult Size. Total length 5′ to 6′, shoulder height to about 3.5′; weight 75 to 300 lbs. (females smaller)

Description. This graceful animal has long, slender legs, large, wide-flaring ears, and a rather long bushy tail. There is a white band around the nose and a dark spot on each side of the chin. The throat, underside of the body, inside of the ears, and underside of the tail are white. In summer the fur of the upper parts is quite chestnut, but in winter the coat is a dense gray or gray-brown. The main "beam" of the male's antlers curves forward, and the tines or "spikes" grow from it. The beam itself is not branched. The antlers are shed after the rutting season and a new set grows the following spring. While growing, the antlers are protected by a fuzzy skin called the "velvet." This is rubbed off in late summer or early fall, when the antlers have stopped growing; they are then hard and polished-looking and are ready for the rutting battles. Full antler growth depends on whether the deer is properly nourished. Theoretically, antlers will grow a little larger each year, but only if the deer gets plenty to eat. Thus, it is not possible to tell the age of a deer by the number of spikes on its antlers. An expert can get this information by examining the amount of wear on the deer's cheek teeth.

Deer are active mostly in late evening, but may be seen at any

Moose

cow

bull

calf

doe

buck

fawn

White-tailed deer

time of the day or night. They can move rapidly—often in a series of great bounds—and they can swim well, and readily take to the water. They often stand on their hind legs, with their front legs braced against a tree trunk, to reach succulent leaves and twigs.

The white underside of the tail sometimes acts as a signal to other deer, for when it is alarmed the deer raises its tail vertically as it leaps off. The resultant flash of white may warn other nearby deer of approaching danger. Adult deer are normally quiet, but if they are scared suddenly they will often utter a loud, snorting shriek. I have frequently been startled by this sound when a hidden deer has suddenly become aware of me. The fawn has a bleating call when trying to draw the attention of its mother.

Breeding. Rutting takes place during the fall and early winter. During this period the bucks (males) develop thick, swollen necks, and search for does (females). Tremendous battles often occur between rival bucks. Once successful mating has been achieved there is a gestation period of about 6.5 months and the doe then gives birth to (usually) two young (fawns) in late May or June. Young does often have only one fawn, while older animals sometimes give birth to triplets. The fawns weigh about four pounds at birth, and are able to walk almost as soon as they are born. Since they are very helpless against predators at this time, they are well camouflaged. Their light orangy coats are dotted with white spots that break up the outline of their bodies when they lie still. As an additional protection, they have little body odor. The fawns stay with their mother until their first winter, by which time the white spots have disappeared.

Habitat. These graceful animals prefer open woodland (mostly deciduous), forest edges, farmland, swamps, and brushy areas.

Food. In winter the twigs and seedlings of maples, apple, birch, oak, willow, witch hazel, sumac, aspen, and wild cherries are favored foods, as well as some blackberry, blueberry, fern, goldenrod, and other low-growing plants. During the rest of the year the leaves and twigs of these plants are still eaten, together with some grasses and other herbaceous plants.

Economic Importance. The white-tailed deer is a most important animal in the Northeast. In the pursuit of their sport, deer hunters spend millions of dollars on equipment, outdoor clothing, travel, and licenses. In some rural areas where winter jobs are scarce, venison still represents a much-appreciated addition to a family's food supply.

Where deer are very common they may do much damage to wooded areas in winter by cleaning out all of the young seedling trees and by girdling saplings. In farming areas or gardens they may ruin many crops.

On some highways they represent a considerable driving hazard at night. If they are feeding at the roadside they may be blinded by the headlights of a fast-moving car and may move out into its path. As well as the deer's being killed, this often results in a serious acci-

dent to the car and its occupants, especially if traffic is fairly heavy.
Comments. Although deer were plentiful in the Northeast during
the early settlement days, they had been mostly killed off by the time
of the American Revolution. Therefore, laws were passed to protect
them from hunting. In some areas no hunting was allowed at all for
many years. Since natural predators such as the wolf and the mountain lion had been eliminated from the Northeast, the deer rapidly
increased in number. Hunting was again permitted, but for many
years hunters were allowed to kill only bucks. These animals are
polygamous, however, and a single buck will breed with several does.
In her normal lifespan (about five years) a doe may produce more
than six young, about half of which will be females that will soon
be adding to the population. Thus, it is easy to see why the Northeast deer population has grown so large. In New York alone there
are an estimated 400,000 deer, and even though approximately
100,000 are shot each year (including some does these days) their
numbers are still too large for parts of the range to support. Thousands starve to death during a bad winter, and thousands more are
so weakened that they have little resistance to disease.

At the present time the annual hunting season seems to be the only
practical way of cutting down the size of the deer herds. Feeding hay
to the deer in winter is not the answer; the strongest animals, which
least need food, drive away the weaker does and fawns until they
themselves have finished eating. In the course of feeding they spoil
much of what is left, so that the rest go hungry.

Range in the Northeast. Throughout the Northeast

Similar Northeastern Species. The moose is much larger, has a mas-
sive, deep body, a tuft of hair hanging from beneath the throat, and
a drooping snout. Bulls have great, palmate antlers.

Moose *(Alces alces)*   PLATE 20

Adult Size. Total length about 9', shoulder height about 6.5'; weight
800 to 1200 lbs., sometimes larger (cows are about 0.25 smaller)

Description. This is the largest member of the deer family in the
world. It has a huge, heavy body, a hump between the shoulders, a
broad, drooping muzzle, and long, relatively slender legs. The tail is
very short. There is a tuft of long hair (the "bell") hanging from the
throat. The animal's color is a very dark brown—sometimes almost
black—with paler legs and some gray on the muzzle. Young calves
are chestnut in color, becoming darker as they get older. Bulls have
enormous, palmate antlers with spikes at their edges. These antlers

may attain a spread of five feet in older animals. Females have no antlers.

Although ungainly in appearance, the moose can move very quietly through the heaviest forest growth. In the open it is capable of short bursts of speed, but normally moves at a walk or a lumbering trot. It swims well, and takes to the water readily. I have seen moose swimming strongly across quite large lakes. Because of its long legs and rather short neck, this animal is forced to kneel when drinking from streams or small pools.

While it prefers to stay on its own for most of the year, the moose will sometimes seek company during the summer, and as many as eight will sometimes be seen together.

Breeding. The rutting season lasts from mid-September until about mid-November. At this time the bulls are very dangerous, since almost any moving object represents a potential rival. Their hoarse bellows can be heard from some distance away, and they thrash the undergrowth with their antlers as they move about. Tremendous battles take place between rival bulls at this time. Once successful mating has taken place there is a gestation period of eight months and the cow then gives birth to one or two calves. Usually the calves are born at the end of May or in early June. A young cow will normally give birth to a single calf, but as she gets older she may have twins or, rarely, triplets. The calves stay with their mothers for their first year, but are driven away to fend for themselves when the female is ready for her next calf.

Habitat. In summer, moose prefer low-lying wet areas such as swamps and bogs, or regions where there are many lakes and streams. When winter comes they often take to higher ground, and may be seen on hillsides and ridges where there are large stands of conifers or mixed woodland.

Moose feeding in a pond

Food. Summer foods consist largely of waterlilies, pondweeds, and other aquatic plants, but this great animal also browses on willows, wild cherry, maple, aspen, blackberries, and a variety of other plant foods. In winter it will also feed on sedges, poplar, birch, alder, and balsam fir.

Economic Importance. The moose is a prized game animal. Its meat tastes somewhat like venison, but is coarser in texture.

Range in the Northeast. Once widespread throughout the Northeast; now found mostly in Maine, especially in the forests near Mt. Katahdin; also seen in New Hampshire and Vermont. During October 1970, a bull moose appeared in New York's Columbia county, and this seems to be the most southern record for the Northeast in many years.

Similar Northeastern Species. The white-tailed deer is much smaller. It is lighter in color and has a much larger tail.

Appendixes

Howbourne

Appendix A

Distribution Tables and Checklists of Animals

TABLE 1

Occurrence of Frogs and Toads in the Northeast (by State)
(may be restricted to certain geographical areas or to local pockets)

	Maine	Vt.	N.H.	Mass.	R.I.	Conn.	N.Y.	N.J.	Pa.	Your List (✓)
E. spadefoot toad				×	×	×	×	×	×	
American toad	×	×	×	×	×	×	×	×	×	
Fowler's toad			×	×	×	×	×	×		
N. cricket frog							×	×		
W. chorus frog							×			
Upland chorus frog								×	×	
N.J. chorus frog							×	×		
Spring peeper	×	×	×	×	×	×	×	×	×	
Gray treefrog	×	×	×	×	×	×	×	×	×	
Bullfrog	×	×	×	×	×	×	×	×	×	
Green frog	×	×	×	×	×	×	×	×	×	
Pickerel frog	×	×	×	×	×	×	×	×	×	
N. leopard (meadow) frog	×	×	×	×	×	×	×	×	×	
S. leopard (meadow) frog					?		×	×		
Mink frog	×	×	×				×			
Wood frog	×	×	×	×	×	×	×	×	×	

TABLE 2

Occurrence of Salamanders in the Northeast (by State)
(may be restricted to certain geographical areas or to local pockets)

× = present

	Maine	Vt.	N.H.	Mass.	R.I.	Conn.	N.Y.	N.J.	Pa.	Your List (√)
Mudpuppy		×	×	×		×	×			
Hellbender							×		×	
Red-spotted newt	×	×	×	×	×	×	×	×	×	
Jefferson's salamander	×	×	×	×	×	×	×	×	×	
Blue-spotted salamander	×	×		×			×	×		
Spotted salamander	×	×	×	×	×	×	×	×	×	
Marbled salamander			×	×	×	×	×	×	×	
E. tiger salamander							×	×		
N. dusky salamander	×	×	×	×	×	×	×	×	×	
Allegheny Mts. salamander							×	×	×	
Red-backed salamander	×	×	×	×	×	×	×	×	×	
Slimy salamander							×	×	×	
Four-toed salamander	×	×	×	×	×	×	×	×	×	
N. spring (purple) salamander	×	×	×	×	×	×	×	×	×	
N. red salamander							×	×	×	
Two-lined salamander	×	×	×	×	×	×	×	×	×	
Long-tailed salamander							×	×	×	

TABLE 3

Occurrence of Turtles in the Northeast (by State)

(may be restricted to certain geographical areas or to local pockets)

$\times$ = present

	Maine	Vt.	N.H.	Mass.	R.I.	Conn.	N.Y.	N.J.	Pa.	Your List (√)
Musk turtle	×	×	×	×	×	×	×	×	×	
E. mud turtle						×	×	×		
Snapping turtle	×	×	×	×	×	×	×	×	×	
Spotted turtle	×		×	×	×	×	×	×	×	
Wood turtle	×	×	×	×	×	×	×	×	×	
Bog turtle						×	×	×		
Blanding's turtle			×	×						
E. box turtle	×		?	×	×	×	×	×	×	
N. diamondback terrapin				×	×	×	×	×		
Map turtle		×					×			
E. painted turtle	×	×	×	×	×	×	×	×	×	
Midland painted turtle		×	×	×	×	×	×	×		
Red-bellied turtle				×						
E. spiny soft-shelled turtle		×					×			

TABLE 4

Occurrence of Snakes in the Northeast (by State)

(may be restricted to certain geographical areas or to local pockets)

✕ = present

	Maine	Vt.	N.H.	Mass.	R.I.	Conn.	N.Y.	N.J.	Pa.	Your List (√)
E. worm snake				✕		✕	✕	✕	✕	
N. ringneck snake	✕	✕	✕	✕	✕	✕	✕	✕	✕	
E. hognose snake			✕	✕	✕	✕	✕	✕	✕	
Smooth green snake	✕	✕	✕	✕	✕	✕	✕	✕	✕	
N. black racer	✕	✕	✕	✕	✕	✕	✕	✕	✕	
Black rat snake		✕		✕	✕	✕	✕	✕	✕	
E. milk snake	✕	✕	✕	✕	✕	✕	✕	✕	✕	
Queen snake							✕			
N. water snake	✕	✕	✕	✕	✕	✕	✕	✕	✕	
N. brown snake	✕	✕	✕	✕	✕	✕	✕	✕	✕	
Red-bellied snake	✕	✕	✕	✕	✕	✕	✕	✕	✕	
Short-headed garter snake							✕			
E. garter snake	✕	✕	✕	✕	✕	✕	✕	✕	✕	
E. ribbon snake	✕	✕	✕	✕	✕	✕	✕	✕	✕	
N. copperhead				✕	✕	✕	✕	✕	✕	
E. massasauga							✕			
Timber rattlesnake		✕	✕	✕	✕	✕	✕	✕	✕	

TABLE 5

Occurrence of Lizards in the Northeast (by State)
(may be restricted to certain geographical areas or to local pockets)

X = present

	Maine	Vt.	N.H.	Mass.	R.I.	Conn.	N.Y.	N.J.	Pa.	Your List (√)
Five-lined skink				X	X	X	X	X		
N. coal skink							X			
N. fence lizard							X	X	X	

TABLE 6

Occurrence of Mammals in the Northeast (by State)
(may be restricted to certain geographical areas or to local pockets)

X = present

	Maine	Vt.	N.H.	Mass.	R.I.	Conn.	N.Y.	N.J.	Pa.	Your List (√)
Opossum	?	X	X	X	X	X	X	X	X	
Masked shrew	X	X	X	X	X	X	X	X	X	
Smoky shrew	X	X	X	X	X	X	X	X	X	
Long-tailed shrew	X	X	X	X			X	X	X	
N. water shrew	X	X	X	X	?	X	X		X	
Pygmy shrew	X	X	X				X			
Least shrew						X	X	X	X	
Short-tailed shrew	X	X	X	X	X	X	X	X	X	
Star-nosed mole	X	X	X	X	X	X	X	X	X	

TABLE 6 (*continued*)

	Maine	Vt.	N.H.	Mass.	R.I.	Conn.	N.Y.	N.J.	Pa.	Your List (√)
E. mole				×	×	×	×	×	×	
Hairy-tailed mole	×	×	×	×	×	×	×	×	×	
Little brown bat	×	×	×	×	×	×	×	×	×	
Small-footed bat	×	×	×	×	×	×	×	×	×	
Indiana bat	?	×	×	×	×	×	×	×	×	
Keen's bat	×	×	×	×	×	×	×	×	×	
Silver-haired bat	×	×	×	×	×	×	×	×	×	
E. pipistrelle	×	×	×	×	×	×	×	×	×	
Big brown bat	×	×	×	×	×	×	×	×	×	
Red bat	×	×	×	×	×	×	×	×	×	
Hoary bat	×	×	×	×	×	×	×	×	×	
Black bear	×	×	×	×			×		×	
Raccoon	×	×	×	×	×	×	×	×	×	
Marten	×	×	×				×			
Fisher	×	×	×				×			
Short-tailed weasel	×	×	×	×	×	×	×	×	×	
Long-tailed weasel	×	×	×	×	×	×	×	×	×	
Least weasel							×		×	
Mink	×	×	×	×	×	×	×	×	×	
River otter	×	×	×	×	×	×	×	?	?	

TABLE 6 (continued)

	Maine	Vt.	N.H.	Mass.	R.I.	Conn.	N.Y.	N.J.	Pa.	Your List (√)
Striped skunk	X	X	X	X	X	X	X	X	X	
Red fox	X	X	X	X	X	X	X	X	X	
Gray fox	X	X	X	X	X	X	X	X	X	
Coyote	?	X	?	X			X		?	
Bobcat	X	X	X	X		X	X	X	X	
Lynx	X	X	X				X			
Mountain lion	?									
Woodchuck (groundhog)	X	X	X	X	X	X	X	X	X	
E. chipmunk	X	X	X	X	X	X	X	X	X	
E. gray squirrel	X	X	X	X	X	X	X	X	X	
Red squirrel	X	X	X	X	X	X	X	X	X	
Fox squirrel							X			
N. flying squirrel	X	X	X	X	?	X	X	X	X	
S. flying squirrel	?	X	X	X	X	X	X	X	X	
Beaver	X	X	X	X	X	X	X	X	X	
Deer mouse	X	X	X	X		X	X	X	X	
White-footed mouse	X	X	X	X	X	X	X	X	X	
E. Woodrat						X	X	X	X	
Bog lemming	X	X	X	X	?	X	X	X	X	
Boreal red-backed mouse	X	X	X	X	?	X	X	X	X	

TABLE 6 *(continued)*

	Maine	Vt.	N.H.	Mass.	R.I.	Conn.	N.Y.	N.J.	Pa.	Your List (√)
Meadow mouse	×	×	×	×	×	×	×	×	×	
Yellow-nosed mouse	×	×	×				×		×	
Pine mouse		×	×	×	×	×	×	×	×	
Muskrat	×	×	×	×	×	×	×	×	×	
Norway rat	×	×	×	×	×	×	×	×	×	
House mouse	×	×	×	×	×	×	×	×	×	
Meadow jumping mouse	×	×	×	×	×	×	×	×	×	
Woodland jumping mouse	×	×	×	×			×	×	×	×	
Porcupine	×	×	×	×		×	×	?	×	
Snowshoe (varying) hare	×	×	×	×	?	×	×	×	×	
E. cottontail				×	×	×	×	×	×	
New England cottontail	×	×	×	×	×	×	×	×	×	
European hare		?	?	×		×	×	?	?	
White-tailed deer	×	×	×	×	×	×	×	×	×	
Moose	×	×	×							

Appendix B

Animal Relationships

Why is a salamander more closely related to a frog than to a jelly-fish? Why are we, as people, more kin to a raccoon than to a turtle? How do we decide just what makes a dog more similar to a weasel than to a woodchuck?

In order to see the relationships among animals—or among plants, too, for that matter—it is important to understand at least the rudiments of just how they are classified. To most people this business of classification may seem very complicated—a mysterious art replete with long, strange-looking names that only a trained biologist could comprehend or would need to know. In actual fact this is not the case at all, and the average person can learn, very quickly, the system that is used today throughout the world. As far as the cumbersome-looking names are concerned, they are merely Latin or Latinized names that help to describe the animal or group of animals, or where they are found. In the old days an animal was often named for a person, but this practice seems to be dying out.

The simplified version of this system of classification—the frame-work of it—is known as the basic hierarchy of classification. What it comes down to is a method where, by a process of elimination, an animal can be placed into each of seven categories that show its rela-tionship to other animals. As an example of how this works, let's consider the spotted salamander, a common enough animal in the Northeast, and see just where it fits into the world of living things.

To begin with, all living things are divided into plants or animals. Each of these two groups is called a kingdom, and this constitutes the first of the seven categories in the basic hierarchy. Everyone will agree, I think, that the spotted salamander is an animal, rather than a plant. So there's our first category:

kingdom: animal.

As we all know, there are huge numbers of different kinds of ani-mals in the world (more than a million): bears, robins, jellyfishes, wasps, whales, and people, to name but a few. So now we really begin our process of elimination by dividing the animal kingdom into a number of major groups called phyla (the singular of the word is phylum). This is the second of the seven categories. Every animal within each of these phyla is more closely related to other animals within its own phylum than it is to animals in other phyla. Micro-scopic animals consisting of but a single cell, such as amoebas and paramecia, are placed into the phylum Protozoa. Animals with jointed legs and having the hard, skeletal parts of their bodies on the outside rather than the inside are all in the phylum Arthropoda (animals such as insects, crabs, and spiders). There are more than twenty of these phyla, another of which includes animals that have backbones and a spinal nerve cord. This is the phylum Chordata. Salamanders have a backbone and a spinal nerve cord, so into the phylum Chordata goes our spotted salamander:

kingdom: animal
phylum: Chordata

Now, a great number of other animals also have backbones, so our next step is to divide the phylum Chordata into smaller groups. Again, all of the animals within each of these groups will be more closely related to each other than to animals in the other groups. This third category is known as the class. One such class in the phylum Chordata consists of animals that have hair somewhere on their bodies, that feed their young milk, and that have various other attributes in common. This is the class Mammalia. Bears, bats, whales, cats, and people belong in this class. Another class in the phylum Chordata is made up of animals that, among other common features, have smooth (no scales, feathers, or hair) usually moist, glandular skins and that typically begin their lives in water and then move out onto land. This is the class Amphibia. The spotted salamander fulfills these criteria, and therefore belongs here. Thus:

kingdom: animal
phylum: Chordata
class: Amphibia

The next step splits the class Amphibia into groups known as orders. One of these orders contains amphibians whose adults have elongated hind legs, a short backbone, and no tails. This is the order Anura—the frogs and toads. Another order within the class Amphibia consists of amphibians that mostly have tails, long, slender bodies and rather short legs—the salamanders. This description certainly fits our spotted salamander, so we place it in this group, the order Caudata. We add that to our growing list of categories:

kingdom: animal
phylum: Chordata
class: Amphibia
order: Caudata

Let's go further. We now divide the order Caudata (the salamanders) into groups of closely related salamanders, based on their anatomy, external features, etc. These smaller groups are known as families. One of them is composed of salamanders with rather chunky bodies, lungs (many salamanders don't have them), and certain other features that set them apart from the salamanders in the other families. The spotted salamander satisfies the requirements for this family, which is known as the family Ambystomidae. So now we have our salamander located in five categories:

kingdom:	animal
phylum:	Chordata
class:	Amphibia
order:	Caudata
family:	Ambystomidae

We continue the process, and divide the family Ambystomidae into smaller groups. All of the salamanders within each of these groups bear a closer resemblance to each other than to those in the other groups. This sixth category is called the genus (the plural of this word is genera). It so happens there are three genera in the family Ambystomidae, one of which is the genus *Ambystoma*. Our spotted salamander has the external features that place it into that genus:

kingdom:	animal
phylum:	Chordata
class:	Amphibia
order:	Caudata
family:	Ambystomidae
genus:	*Ambystoma*

Now we're almost there. Only one step remains, and that is to designate the actual *kind* of salamander within the genus *Ambystoma* —the animal that will normally breed only with others like itself. For this final category we use the term species (the plural is the same word). The species (specific) name of the spotted salamander is *maculatum*. There we have it:

kingdom:	animal
phylum:	Chordata
class:	Amphibia
order:	Caudata
family:	Ambystomidae
genus:	*Ambystoma*
species:	*maculatum*

The last two categories, genus and species, are used as the scientific name of the animal: the spotted salamander *(Ambystoma maculatum)*.

Why bother with a scientific name? Why not be content with the common name, spotted salamander? There is a very good reason: common names vary! Depending upon where an animal is found, people may have their own names for it. Near the Okefenokee Swamp, in Georgia, the bird most people know as the wood stork is often called the hammerhead. In some areas, the well-known timber rattlesnake is known as the banded rattlesnake. On the other hand, two different animals may have the same common name. A good example of this is in the case of the robin. There is a small bird found in Britain, Europe, and North Africa that has a red breast and that the British call a robin. When some of the early settlers from Britain came to North America they saw a bird that was much bigger than their robin, and that had an orangy, rather than a red, breast. Nevertheless, it reminded them of the robins they had left behind, so they called this new bird a robin. Now here are two completely different *species* of birds with the same common name, obviously a confusing situation. The addition of the scientific name at once ends this confusion. The scientific name is an international name. A scientist or an interested layman can discover at once that *Erithacus rubecula* refers to the robin found in Britain, Europe, and North Africa, while *Turdus migratorius* identifies the American robin. It doesn't matter whether the person doing the investigating lives in Britain, Europe, Africa, the United States, India, or Mongolia; the scientific name remains the same. You will notice that the generic name is always capitalized, while the specific name is not: the spotted salamander *(Ambystoma maculatum),* the black bear *(Ursus americanus),* the bullfrog *(Rana catesbeiana),* and so on.

Sometimes you will see a third word tacked on to the end of a scientific name, for example, *Chrysemys picta picta.* This last name is merely the subspecies. It is used where a species has been divided into recognizably different groups, and it is usually of importance only to a scientist. *Chrysemys picta picta* (usually written as *Chrysemys p. picta*) is the eastern painted turtle. *Chrysemys picta marginata* is the midland painted turtle. These two turtles will interbreed where their ranges overlap, but are otherwise separated geographically.

One more example. Let's see where the northern copperhead belongs:

kingdom:	animal
phylum:	Chordata (has a backbone, etc.)
class:	Reptilia (scaly skin, eggs laid on land, etc.)
order:	Squamata (snakes and lizards)
family:	Crotalidae (pit vipers)
genus:	*Agkistrodon* (copperheads and water moccasins)
species:	*contortrix*
subspecies	*mokeson*

Thus, the scientific name of the northern copperhead is *Agkistrodon contortrix mokeson.*

Glossary

ALTRICIAL   young that remain in the nest for a prolonged period after birth, and therefore require much parental care (e.g., mice, rabbits)

ARBOREAL   adapted for living mostly in trees (e.g., flying squirrels)

BEAM   the main section of a deer's antler, from which the tines (spikes) arise

BOG   where a freshwater pond has become partially or entirely filled by organic matter from the growth and decay of water plants

BRIDGE   the lateral section of a turtle's shell that connects the carapace to the plastron

BROWSING   feeding upon leaves and twigs of trees and shrubs (e.g., deer)

CARAPACE   the upper part of a turtle's shell

CANINE TEETH   four long, pointed teeth found in the front of the jaw of many mammals. In many members of the order Carnivora, these teeth are used for stabbing and holding prey

CARNIVOROUS   Meat eating (not necessarily a member of the order Carnivora; sharks, snakes, and many other animals are also carnivorous)

CARRION   dead animal matter

CLASS   a category used in classifying animals (see Appendix B, "Animal Relationships")

CONIFEROUS   applied essentially to evergreen trees such as pines, spruces, firs, and hemlocks, but also to some shrubs

CUD   in deer (and cattle, sheep, and some other large plant-eaters), where it is dangerous for them to expose themselves for long periods, the food is swallowed rapidly. It passes into the first section (rumen) of a specialized stomach made up of four compartments. Later, when the animal feels safe, it regurgitates small masses (cuds) of this food, chews them thoroughly, and reswallows them. This time the food passes through the other compartments (reticulum, omasum, and abomasum), where more complete digestion takes place. Cud-chewers are known as ruminants (from their possession of a rumen).

Deciduous applied to trees and shrubs that lose all of their leaves in the fall (e.g., maples, oaks, viburnums)

Dormancy a period when an animal sleeps for a prolonged time, but where the body metabolism does not slow down as in hibernation. Bears, skunks, raccoons, and squirrels are examples of animals that become dormant in winter

Dorsal referring to an animal's back (or sometimes its entire upper part)

Dorso-lateral fold found in some frogs (e.g., green frog), a thickened ridge running along the edge of the back from behind each eye

Ectothermic cold-blooded, i.e., where the temperature of the body fluctuates depending upon the environment. Amphibians and reptiles are ectothermic. Mammals are warm-blooded (endothermic); their temperature is normally unaffected by their environment

Eft the immature, land-dwelling stage of a newt

Evergreen applied to trees and shrubs that lose their needles or leaves only a few at a time, so that they are basically green throughout the year (e.g., spruces, pines, firs, yews)

Family a category used in classifying animals (see Appendix B, "Animal Relationships")

Fang a specialized tooth that, in the three Northeastern poisonous snakes, is long, sharp, and hollow, and through which the venom is injected (found *only* in venomous snakes)

Form a shallow depression in grass or soil in which a hare habitually lies when not active

Genus a category used in classifying animals (see Appendix B, "Animal Relationships")

Gestation the period from fertilization to birth; pregnancy

Girdling the removal of bark completely around the stem or branch of a woody plant. When this occurs, the plant dies above that point. (Porcupines, mice, deer, and rabbits are examples of mammals that often girdle plants)

Grazing feeding upon grasses and herbs (e.g., woodchucks)

Habitat the kind of place in which an animal prefers to live (e.g., woods, swamps), not the geographic location

Herbaceous applied to most plants that are not woody. Grasses, dandelions, wildflowers, etc., are herbaceous; trees and shrubs are not (nor are fungi, mosses, and other primitive plants)

Herbivorous plant eating (e.g., rabbits, deer, woodchucks)

Home range the area in which an individual animal or family group is active. (If this area is defended against animals of the same species it is known as a territory)

Incisor teeth simple teeth in the front of the jaw that are adapted for cutting or gnawing

INTERFEMORAL MEMBRANE   a membrane enclosing or partially enclosing the tail in many bats

KEELED SCALE   a scale that has a slight ridge along its center. In some snakes (e.g., the northern water snake) these keels are very obvious, and give the skin a very rough appearance; in others (e.g., black rat snake) they are difficult to see, and merely look like a fine line along the center of the scale. Many snakes have no keeled scales at all, so that the presence or absence of keeled scales is sometimes important in identification.

LARVA   as used in this book, the immature stage between the egg and the adult of an amphibian (also known as a tadpole in the case of frogs and toads)

LEVERET   the young of a hare

MARSH   a watery area standing at or close to ground level over a wide area, in which grow sedges, reeds, grasses, and broad-leafed herbaceous plants

MIGRATION   the annual travels of animals to and from their breeding grounds or seasonal feeding areas

OMNIVOROUS   feeding upon both plant and animal matter (e.g., black bear, raccoon, Norway rat, opossum, man)

ORDER   a category used in classifying animals (see Appendix B, "Animal Relationships")

PALMATE   shaped like a hand with fingers extended (e.g., moose antlers)

PAROTOID GLAND   a lumplike gland that is usually quite prominent in toads, and that is located behind the eye and slightly above the round tympanum (ear membrane). These glands help to protect the toad by secreting a substance that is very distasteful to other animals

PHYLUM   a category used in classifying animals (see Appendix B, "Animal Relationships")

PIT VIPER   all three of the Northeastern poisonous snakes are in the pit viper family (Crotalidae), so-called because of a small opening (the pit) located near each nostril and acting as a heat-receptor

PLASTRON   that part of a turtle's shell located on the underside of the animal

PRECOCIAL   young that are born with eyes open and fur present, able to move about shortly after birth (e.g., hares, deer)

PREDATOR   an animal that preys upon other animals (e.g., snakes, weasels)

PREHENSILE   capable of grasping (e.g., an opossum's tail)

RANGE   the geographic distribution of a species (See also Home Range, something entirely different)

RUTTING SEASON   the time of sexual activity in species such as deer

SCUTE   a large plate or scale, seen on turtles' shells and on the undersides of snakes

SEDGE a rushlike or grasslike plant usually growing in clumps in or near shallow water (many species)

SPECIES a category used in classifying animals (see Appendix B, "Animal Relationships")

SPERMATOPHORE a small, stalked capsule filled with live sperm that is deposited by a male salamander and later picked up by a female and taken into her body to fertilize the eggs

SWAMP Like a marsh, this is a watery area, but it contains trees and shrubs as well as herbaceous plants

TERRESTRIAL living on the ground, rather than being aquatic or arboreal

TUSSOCK a clump of grasses or sedges

VIXEN a female fox

WEAN to accustom a young mammal to food other than its mother's milk

Further Reading

One of my firmest beliefs is that the best way to learn about animals is to get out into the field and seek them out for oneself. Knowledge obtained in this way is knowledge that sticks, and personal observations and discoveries add immeasurably to one's enjoyment of the countryside. Nevertheless, there are times when one may wish to check out certain items or just plain read about certain animals.

The references that follow are obviously not a complete listing. They are merely books that will foster the reader's interest. In turn, they will act as signposts to further reading. (Don't forget the periodicals! Magazines such as *Audubon Magazine* and *Natural History,* and state publications such as *The Conservationist,* published by the State of New York Department of Environmental Conservation, often contain valuable information on animals of the Northeast.)

Amphibians and Reptiles

Bishop, Sherman C. 1962. *Handbook of Salamanders.* New York: Hafner Publishing Co.

Carr, Archie F., Jr. 1952. *Handbook of Turtles.* Ithaca: Cornell University Press.

Ditmars, Raymond L. 1952. *A Field Book of North American Snakes.* Garden City: Doubleday & Co., Inc.

Goin, Coleman J., and Goin, Olive B. 1962. *Introduction to Herpetology.* San Francisco: W. H. Freeman and Co.

Oliver, James A. 1955. *The Natural History of North American Amphibians and Reptiles.* Princeton: D. Van Nostrand Co.

———. 1958. *Snakes in Fact and Fiction.* New York: The Macmillan Co.

Pope, Clifford H. 1944. *Snakes Alive and How They Live.* New York: The Viking Press.

———. 1949. *Turtles of the United States and Canada.* New York: Alfred A. Knopf.

Smith, Hobart M. 1946. *Handbook of Lizards*. Ithaca: Comstock Publishing Co., Inc.

Wright, Albert Hazen, and Wright, Anna Allen. 1949. *Handbook of Frogs and Toads*. Ithaca: Comstock Publishing Co., Inc.

Mammals

Anthony, H. E. 1928. *Field Book of North American Mammals*. New York: G. P. Putnam's Sons.

Cahalane, Victor H. 1961. *Mammals of North America*. New York: The Macmillan Co.

Drimmer, Frederick, ed. 1954. *The Animal Kingdom*. 3 vols. New York: The Greystone Press.

Godin, Alfred J. 1977. *Wild Mammals of New England*. Baltimore: The Johns Hopkins University Press.

Hamilton, William J., Jr. 1943. *The Mammals of Eastern United States*. Ithaca: Comstock Publishing Co., Inc.

Walker, E. P. 1975. *Mammals of the World*. 3rd ed. rev. by John L. Paradiso. 2 vols. Baltimore: The Johns Hopkins University Press.

General

Allen, Durward L. 1962. rev. ed. *Our Wildlife Legacy*. New York: Funk and Wagnalls Co., Inc.

Martin, Alexander C.; Zim, Herbert S.; and Nelson, Arnold. 1961. *American Wildlife and Plants: A Guide to Wildlife Food Habits*. New York: Dover Publications, Inc.

Matthiessen, Peter. 1959. *Wildlife in America*. New York: The Viking Press.

About the Author and the Illustrator

KENNETH A. CHAMBERS, author and editor of numerous publications on wildlife, is Lecturer in Zoology at the American Museum of Natural History in New York City. Besides his lecturing and extensive field work, he has led wildlife study tours for the museum in the eastern United States and Alaska. He is a fellow of The Explorers Club. Degrees in wildlife management (Cornell) and adult education (C.U.N.Y.) followed a farming apprenticeship in his native England. He now divides his time between his work in New York City and his farm near New Lebanon, N.Y.

H. WAYNE TRIMM is one of America's leading illustrators of wildlife. He serves as art director for *The Conservationist,* a magazine published by the New York State Department of Environmental Conservation, and teaches nature painting and drawing at the State University of New York at Albany and the College of Saint Rose. He lives in Albany and has a farm not far from Kenneth Chambers.

THE JOHNS HOPKINS UNIVERSITY PRESS

This book was composed in Linotype Baskerville text with Palatino semibold display type by Maryland Linotype Composition Company, Inc., from a design by Alan Carter. It was printed on 80-lb. Paloma Matte and bound in cloth by Universal Lithographers, Inc.

Library of Congress Cataloging in Publication Data

Chambers, Kenneth A
  A country-lover's guide to wildlife.

  Bibliography: P. 226
  Includes index.
  1. Mammals—Northeastern States. 2. Reptiles—
Northeastern States. 3. Amphibians—Northeastern
States. 4. Wild animals as pets. I. Title.
QL719.N92C48   596'.0974   79-4338
ISBN 0-8018-2207-6